P9-AQW-754

SAXON MATH

Course 3

Stephen Hake

T-S
Ma-Sa98
Ha-8-tr
v. 3

Graphing Calculator Activities

SAXON

A Harcourt Achieve Imprint

www.SaxonPublishers.com
1-800-284-7019

TI-84™ is a trademark of Texas Instruments.

TI-83 Plus™ is a trademark of Texas Instruments.

TI-73 Explorer™ is a trademark of Texas Instruments.

TI-Presenter™ is a trademark of Texas Instruments.

ViewScreen™ is a trademark of Texas Instruments.

© 2007 Harcourt Achieve Inc. and Stephen Hake

Saxon is a trademark of Harcourt Achieve Inc.

ISBN: 1-5914-1979-4

Printed in the United States of America

Manufacturing Code: 2 3 4 5 6 7 8 9 202 13 12 11 10 09 08 07

Teacher Resources for Graphing Calculator Activities

Graphing Calculator Activities

About the Graphing Calculator Activities

At Saxon Publishers, we believe that people learn by doing. Students learn mathematics not by watching or listening to others, but by doing the problems themselves. The graphing calculator activities in this book are designed for whole-class instruction that leads into individual and group practice. However, students could do most of each activity individually, following the demonstration steps and completing the practice problems.

Technology in the Classroom

Integration of technology into any curriculum is important for students to be competitive in a constantly changing academic landscape. This is even more important for students in math and science. Technology, like the graphing calculator, can be a helpful tool that allows students to step away from the arithmetic and explore a deeper level of mathematics. For instance, **Graphing Calculator Activity 2** demonstrates graphing points on a coordinate plane, which students do by hand in Investigation 1. But practicing it on the calculator will lead some students to find patterns in how data is written in table format and reinforce the order in graphing coordinates. This is especially true for students with special needs, as the calculator may liberate them from challenges in mathematical calculation to a higher level of understanding.

Teacher Resources

Each graphing calculator activity has a teacher resource section, which lists the calculator functions, materials, demonstration tips, practice tips, and the answer key for each activity. Preview each activity and its teacher resource section before using the handout in your lesson plan. You may want to go through each demonstration on your own so that you can prepare for the time requirement. A graphing calculator activity may seem time consuming because of its number of pages. However, the demonstration may actually take up less time than the activities with fewer pages.

Teacher Resource

Methods

The most effective method of teaching the calculator demonstrations is to guide your class through the steps. Pairing students may help them to stay on track. It is recommended that you display the keystrokes with a graphing calculator display, such as the ViewScreen Panel™ or TI-Presenter™. You may also find it helpful to have a large poster or a transparency of the calculator to point to as you press buttons during the demonstration.

The most effective method for learning the functions on the graphing calculator is to use the demonstration steps as a guide for the first practice problem. Then students should try to solve the rest of the practice problems without looking at the demonstration steps. This will help them commit to memory the functions needed in the graphing calculator to solve that particular type of problem.

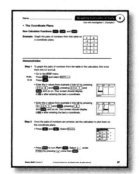

Student Resource

The most effective method for retention is for students to keep a binder of the demonstrations as a reference guide. Students can also make study cards with the keystrokes used in the demonstration steps and keep them bound together for quick reference. You may occasionally allow students to verify a problem from the written practice with the graphing calculator.

Demonstration Tips

Most of the demonstrations are written to the **TI-83+** graphing calculator.
However, there are times when separate instructions are written for both
the TI-83+ and **TI-73.** The illustrations of the calculator screens are mostly
of the TI-83+, unless otherwise noted. The TI-73 does have similar screen
displays but may have slight differences from the TI-83+. Also note that
the keystrokes for the TI-83+ are 100% compatible with the **TI-84.**

- To select a menu function, students can press the number next to it or press the
  down-arrow key and enter. These keystrokes are in addition to the keystrokes
  listed in the demonstration.

  For example, to select **LinReg(ax+b)** press ⬚4 or press ⬚ three times
  and ⬚ENTER.

- If students lose their way in a demonstration, you can instruct them to press ⬚2nd
  and ⬚MODE to return to the **HOME** screen and start over.
- To erase the display on the **HOME** screen, students can press ⬚CLEAR.
- If the settings on the calculator are vastly different from a demonstration, try
  resetting them.

  TI-83+ Press ⬚2nd ⬚+. Then select ⸢Reset...⸣, ⸢Defaults...⸣, and
  ⸢Reset⸣. The screen should display **Defaults set**. If not, try pressing
  ⬚2nd and then holding down ⬚.

  TI-73 Press ⬚2nd ⬚0. Then select ⸢Reset...⸣, ⸢Defaults...⸣, and
  ⸢Reset⸣. The screen should display **Defaults set**. If not, try pressing
  ⬚2nd and then holding down ⬚.

- If the settings on the calculator are corrupted or the functions are not behaving as
  they are supposed to, try resetting the memory.

  TI-83+ Press ⬚2nd ⬚+ and select ⸢Reset...⸣. Press ⬚ twice and select
  ⸢All Memory...⸣ then ⸢Reset⸣. The screen should display
  Resetting All... then **Mem cleared**. If not, try pressing ⬚2nd and then
  holding down ⬚.

  TI-73 Press ⬚2nd ⬚0 and select ⸢Reset...⸣, ⸢All RAM...⸣, then⸢Reset⸣.
  The screen should display **RAM cleared**. If not, try pressing ⬚2nd and
  then holding down ⬚.

Practice Tips

- You may choose to do only the demonstration of the graphing calculator activity, as time permits. However, the practice problems will help students apply the calculator functions to real-world and authentic situations.

- Some of the practice problems require group activity. Look ahead to each practice set and determine if a transition from the demonstration to group activity is needed. Any materials needed are listed in the teacher materials section for each graphing calculator activity.

- An effective method for students to complete the practice problems is to have students work individually and then pair students to share how they got the solution. You may also want a volunteer student to demonstrate a solution on a graphing calculator for the class.

Measures of Central Tendency

TI-73/TI-83+ Functions

 STAT and **LIST**

Materials

- Teacher-provided materials: TI-83+ or TI-73 graphing calculators, number cubes from the manipulative kit

Demonstration Tips

- You may choose to use the demonstration to reinforce the solution to example 5 in Lesson 7.

- To select a menu function, students can press the number next to it or press the down-arrow key and enter. These keystrokes are in addition to the keystrokes listed in the demonstration.

- If **L1** is not clear before students enter the data points, instruct them to first go to the **HOME** screen by pressing **2nd** and **MODE**.

  TI-83+ Then have the students press **2nd** **+**, select ClrAllLists, and then press **ENTER**. The screen should display **Done**.

  TI-73 Then have the students press **2nd** **0**, select ClrAllLists, and then press **ENTER**. The screen should display **Done**.

- The **MATH** submenus for the TI-83+ and TI-73 may look very different from each other, but mean(and median(are in the same location.

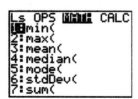

TI-83+   TI-73

- Students may forget to go to the **HOME** screen when calculating the mean, median, or mode. The calculation will be placed in **L1** and may affect their calculations for the other measures of central tendency. Instruct those students to press [DEL] for the calculation in the list and [2nd] [MODE] to go to the **HOME** screen.

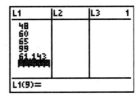

Practice Tips

- For problem 1, you may suggest that the students enter their sums into the calculator as they roll the number cubes.
- Discuss with students how the mean in problem 2 would be a less accurate measure of central tendency than the median. **The mean is a lot higher than the median, but the median is more like most of the numbers**.

Answers

Step 2: mean = 189.8333333

Step 3: median = 186

Step 4: mode = 175

1. **a.** Sample: 6.56

   b. Sample: 6

   c. Sample: 6

   d. mode; Sample: The mode best represents the typical sum rolled because it represents the number that occurs most frequently in the data set.

2. **a.** 1001.666667

   b. 620

   c. 620

   d. mean; The mean would distort this fact because it is over 1001 and the median is 620.

The Coordinate Plane

TI-73/TI-83+ Functions

PLOT, **ZOOM**, and **TRACE**

Materials

- Teacher-provided materials: TI-83+ or TI-73 graphing calculators, graph paper

Demonstration Tips

- This demonstration may be used in addition to or in place of example 2 in Investigation 1.

- If **L1** and **L2** are not clear before the students enter the data points, instruct them to go to the **HOME** screen by pressing **2nd** and **MODE**.

  TI-83+ Then have the students press **2nd** **+**, select **ClrAllLists**, and then press **ENTER**. The screen should display **Done**.

  TI-73 Then have the students press **2nd** **0**, select **ClrAllLists**, and then press **ENTER**. The screen should display **Done**.

- When setting the calculator to plot the pairs of numbers on the coordinate plane, you may want to explain the connection between **L1** and the x-coordinates and between **L2** and the y-coordinates. You may also explain that the values in the first column/row in any table are usually entered into **L1** and explain that the values in the second column/row are entered into **L2**.

Work Hours	Total Pay
2	20
4	40
6	60
8	80

Average Speed	20	35	50	65
Miles per Gallon	10	15	21	17

- Students often press GRAPH to plot the pairs of numbers. Although the resulting graph may be correct, this function key is not always reliable. Instead, students should press ZOOM and select ⟦ZStandard⟧ when they generate a graph. The graphs below show what can happen when students use different **ZOOM** functions instead of ⟦ZStandard⟧.

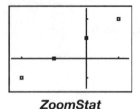

ZoomStat

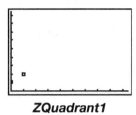

ZQuadrant1

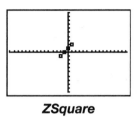

ZSquare

Practice Tips

- For problem 1, remind the students to call out the coordinates of each vertex one at a time to their partners. Also make sure that they enter their partners' *x*-values in **L1** and *y*-values in **L2**.
- Students may have difficulty understanding the scenario for problem 2. Draw a diagram on the board of the airplane's path and discuss the meaning of the *x*- and *y*-values.

Answers

Step 3: The graph of the pairs of numbers should look like the given figure.

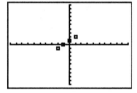

1. See students' work to verify that the pairs of numbers were entered correctly into **L1** and **L2** in the calculator.

    a. Sample: If either student made a simple object with few vertices, then the other student could quickly guess it.

    b. Sample: Make an object with more vertices.

    c. Sample: Yes. It took longer to guess each other's object because we each made objects with many vertices.

2. Choice B

Saxon Math Course 3

Adding and Subtracting Mixed Numbers

TI-73/TI-83+ Functions

menu:Frac or (%)

Materials

- Teacher-provided materials: TI-83+ or TI-73 graphing calculators

Demonstration Tips

- You may choose to use the demonstration to reinforce the solution to example 5 in Lesson 13 rather than replacing it.

- Review with students why $92 + \frac{5}{8} = 92\frac{5}{8}$.

- Discuss why the use of parentheses, especially in subtraction, is necessary with mixed numbers. If needed, have the students explore a subtraction problem with and without parentheses.

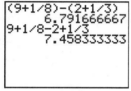
TI-83+

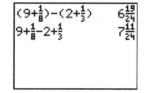
TI-73

- If students using a TI-83+ find it confusing to use fPart when separating the whole and decimal parts, have them subtract the whole number from the answer and then press ENTER. The screen should display just the decimal part of the number.

- Remind students using a TI-73 to use (%) when entering a fraction. However, if they follow the TI-83+ instructions, they can convert the decimal number to a mixed number by pressing (F↔D) and then ENTER.

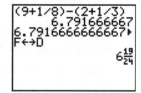

Practice Tips

- If you cannot divide the students into groups of three, tell them to work problem 1 as if there are 3 students in the group.

- Discuss the real-world value of the answer to problem 2. $\frac{69}{80}$ **is not standard measure, but $\frac{70}{80}$ or $\frac{7}{8}$ is standard.**

- Ask students what weight is printed on the box of cereal. **Net weight, which is the cereal without the packaging.**

Answers

Steps 1–3: $77\frac{1}{8}$ inches

1. **a.** 4 cups of pasta; $\frac{9}{16}$ pounds of chicken; $2\frac{1}{4}$ ounces of feta

   b. $5\frac{1}{3}$ Tbsp. of olive oil; $\frac{1}{2}$ Tbsp. of Italian Seasoning

   c. $2\frac{1}{6}$ cups of vegetables

2. $3\frac{69}{80}$ pounds

3. Choice D

Powers and Roots

TI-73/TI-83+ Functions

 and

Materials

- Teacher-provided materials: TI-83+ or TI-73 graphing calculators

Demonstration Tips

- This demonstration can be used in addition to or in place of example 5 in Lesson 15.

- Discuss how the calculations of the first and second bullets of each step are inverse operations of each other.

- Some students may confuse **xth** and **nth** root terminology. Discuss how mathematical notation may be different on the calculator from that in the book.

- The **power** function is sometimes referred to as a **caret**.

- Students may discover that they do not need to use in calculating the root of a number. Encourage them to use the parentheses for proper form.

Practice Tips

- Students may need help in finding the dimensions of the square in problem 1. Ask them which operation is used to find the area of a square. **Squaring the side length.** Then ask which inverse operation "undoes" the area of a square. **Taking its square root.**

- Be aware of students who mistakenly take the square root of the area for the rectangular sheet. Discuss why taking the square root of a rectangular area is not an inverse operation. **The area of a rectangle is not one side length squared but two different side lengths multiplied together.**

- Some students may need help understanding that the decrease on one dimension is added to the other dimension. Discuss what the dimensions would be for a 10 by 10 square if 2 units were taken from one side and the requirement was followed. **12 by 8 units**

- For problem 2, remind students that cannot use the calculator to take the square root of 16,200.

- In problem 3, tell students who are having difficulty to simplify each expression as a solution strategy.

Answers

Step 1: 12

Step 2: 3; The students should have on their calculator screens what is shown in the given figure.

```
³√(27)
             3
Ans³
            27
```

 Saxon Math Course 3

Step 3: 2; The students should have on their calculator screens what is shown in the given figure.

```
5×√32
              2
Ans^5
             32
```

1. **a.** 35 in. by 35 in.

   b. 45 inches by 25 inches

   c. No, because the rectangular area is smaller than the square area but is the same price.

2. She could square 127 to see the answer is close to 16200. Her calculation is correct because $127^2 = 16,129$.

3. Choice C

Multiplying and Dividing Mixed Numbers

TI-73/TI-83+ Functions

[frac] or (A⅛↔⅝)

Materials

- Teacher-provided materials: TI-83+ or TI-73 graphing calculators, square- or rectangular-shaped objects

Demonstration Tips

- You may choose to use the demonstration to reinforce the solution to example 1 in Lesson 23 rather than replacing it.

- For steps 1 and 2, enter them as an addition of the whole number part to the fraction part instead of converting the mixed fractions to improper fractions, (as in **Graphing Calculator Activity 3**). However, doing so will skip the skill of entering an improper fraction in the calculator.

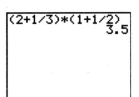

- In step 3, students using a TI-83+ may try to convert 3.5. This will result in an improper fraction, which students will need to convert to a mixed fraction by hand.

- You can demonstrate example 1b on the graphing calculator using steps 1–3. Remember to use the specific mixed numbers to the example and to use ⟨ ÷ ⟩ instead of ⟨ × ⟩ in step 2.

Practice Tips

- For problem 1, instruct students to measure to the nearest $\frac{1}{16}$ inch. Remind them that area is length × width.

- Some students may have difficulty with problem 2. Point out the mixed numbers $2\frac{2}{3}$ and $2\frac{1}{4}$ in the problem. Then ask what they think needs to be done to the amount of milk to calculate all that is used by the family. **Multiply it by the factor used by the family.**

- Some students may have difficulty seeing the division in problem 3. Some will want to multiply the fractions. Discuss why the product is not a real-world answer. **The liquid would weigh more than the cup and liquid combined.**

Answers

Steps 1–3: for $2\frac{1}{3} \times 1\frac{1}{2} = 3\frac{1}{2}$; The students should have the following displayed on their screens.

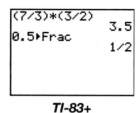

TI-83+

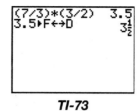

TI-73

Steps 1–3: for $1\frac{2}{3} \div 2\frac{1}{2} = \frac{2}{3}$; The students should have the following displayed on their screens.

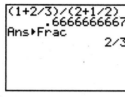

TI-83+

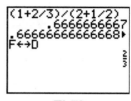

TI-73

1. Check students' work for correct measurement and calculation of area.

   a. Check students' work for correct improper fraction form. Sample:

   ```
   (117/16)*(15/4)
              27.421875
   Ans▶Frac
                1755/64
   ```

   b. Check students' work to verify that the final answer is converted to a mixed number. Sample:

   ```
   (7+5/16)/(4+1/8)
           1.772727273
   Ans−1
            .7727272727
   Ans▶Frac
                  17/22
   ```

   c. Sample: No, because the denominator of the fraction is not measurable on the ruler.

2. 6 cups

3. $\frac{2}{5}$ pounds

Scientific Notation of Large Numbers

TI-73/TI-83+ Functions

Sci and EE

Materials

- Teacher-provided materials: TI-83+ or TI-73 graphing calculators, adding machine tape

Demonstration Tips

- You may choose to use the demonstration to reinforce the solutions to examples 1 and 2 in Lesson 28 instead of replacing them.

- Students may use ⟨,⟩ when entering large numbers. This will result in an error message. Tell them to select Quit and reenter the number without commas.

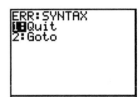

- The **MODE** menus for the TI-83+ and TI-73 may look very different from each other, but Sci is the same place.

         TI-83+   **TI-73**

- If the students need to confirm that **E** symbolizes **×10** in scientific notation, have them use the form in the calculator by using 9.3, ×, 10, ^, and 7. Then press ENTER.

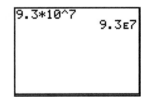

- Remind students that ^ denotes **power**.

- For step 2, have the students press ⟨1⟩ ⟨.⟩ ⟨5⟩ ⟨×⟩ ⟨1⟩ ⟨0⟩ ⟨^⟩ ⟨6⟩ ENTER.

Saxon Math Course 3

Practice Tips

- For problem 1, you may want to use the board instead of adding machine tape. The numbers to call out are listed to the right. Call the numbers out in order one at a time and encourage students to use the **E** function on the calculator. Remind students to switch from **Normal** to **Scientific** modes when necessary.

8.6E2
5.42E5
1.024E9
157,000
24,000
6,987,000,000
3.14E3
45,000

- Encourage students to use the calculator functions to find the answer to problem 2.

- One strategy for solving problem 3 would be to convert each standard number to scientific notation in the calculator. If none of the choices match the scientific number in the problem, then the answer would be choice D.

- Before students try to convert each number in the inequality to standard form, encourage them to mentally convert the numbers based on the power of 10.

Answers

Step 1: 9.3×10^7

Step 2: 1.5×10^6; 1,500,000

1. Check students' work to verify accurate positioning on the number line.

   a. Sample: Not knowing all of the numbers before hand. As some of the numbers were called out, we had to erase and reposition the numbers we had already placed.

   b. Sample: 1,024,000,000; 6,987,000,000; First we had to reposition with 1,024,000,000 because it was very large. But then we had to reposition again because 6,987,000,000 became the highest number.

   c. Sample: No, because they were close to other numbers that were already correctly positioned.

2. $\$5.15 \times 10^{10}$

3. Choice B

4. No, the inequality symbol should be >. Any number with a larger power is greater than a number with a smaller power.

Subtracting Integers

TI-73/TI-83+ Function

Materials

- Teacher-provided materials: TI-83+ or TI-73 graphing calculators, **NUMLINE** application (for TI-73)

Demonstration Tips

- This demonstration can be used in addition to or in place of example 1 in Lesson 33.

- Notice that we do not use the **addition** function for (+2) in step 1. Explain to students that the calculator assumes an integer is positive unless the **negative** function is in front of it. Discuss how this is also true in algebra.

- Some students may discover that ▆▆ and ▆▆ are not necessary. However, encourage students to use the parentheses for separation of operators and values.

```
-3--2
            -1
```

- Students may use ▆▆ instead of (−). This will result in an error message. Tell students to select Quit and reenter the numbers using the **subtraction** and **negative** functions correctly.

```
ERR:SYNTAX
1:Quit
2:Goto
```

Practice Tips

- For problem 1, tell students to draw the number line at the top of the paper and to draw the arrows below the number line as you call out the numbers. Read the numbers in the provided list one at a time so that students can calculate the values and draw the arrows on their number lines.

```
0+6
+6
−20
−10
−2
−(−10)
−(−20)
−6
−6
```

Saxon Math Course 3

- Students using a TI-73 can use the **NUMLINE** application in addition to or in place of the hand-drawn number line. Tell the students to press [APPS] and select [NUMLINE]. After pressing any key, select [NumberLine]. As you call out the integers, the students can enter them into the calculator and press [ENTER] to see their movement on the number line.

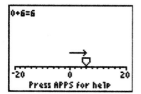
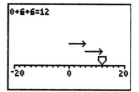

- Even though they can calculate the answers to problem 2 with mental math, encourage students to use the calculator functions.

Answers

Step 1: -5

Step 2: -1

1. Check students' work to see if they are drawing the addition and subtraction with arrows correctly.

   a. 0, the number we started out with

   b. We subtract the same numbers that we add.

   c. Sample: Yes. We would have to make sure that the total amount we added is subtracted with the different integers.

2. **a.** $35000 - (-3500)$

   b. $35000 + (+3500)$

   c. Choice D

3. Yes, because adding a positive is the same as subtracting a negative.

Multiplying and Dividing Integers

TI-73/TI-83+ Function

Materials

- Teacher-provided materials: TI-83+ or TI-73 graphing calculators, standard deck of cards

Demonstration Tips

- This demonstration can be used in addition to or in place of example 1 in Lesson 36.

- In step 1, parentheses are used to indicate multiplication. Students may discover that they can multiply using 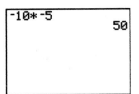 instead and get the same answer.

- Students often use ⬛ instead of $(-)$. This will result in an error message. Tell students to select Quit and reenter the numbers using both the **subtraction** and **negative** functions correctly.

- Notice we do not use the **addition** function for (+10). Explain to students that the calculator assumes an integer is positive unless the **negative** function is in front of it. Discuss how this is also true in algebra.

- Encourage students to use the parentheses for separation of operators and values, especially in problems involving division, such as in step 2.

- For step 2, students with a TI-73 can follow the instructions for the TI-83+ and get the same answer. However, they will not be practicing division as stacked fractions.

Practice Tips

- Students may have difficulty understanding the rules in problem 1. Use the figures below as a sample of cards drawn and the arithmetic that should be displayed on the calculator.

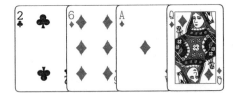

Black 2   Red 6   Red Ace   Red Queen

- Discuss each choice in problem 2 in mathematical and real-world terms. For instance, discuss why a negative number is usually not associated with income.

Saxon Math Course 3

- If students are having trouble without a calculator in problem 3, have them multiply the signs alone. Write the following on the board:

$$(-)(+)(+)(-)(-)$$
$$(-)(+)(-)(-)$$
$$(-)(-)(-)$$
$$(+)(-)$$
$$(-)$$

Answers

Step 1: +50; −50

Step 2: +2; −2

1. Check students' work for the multiplication and division of the integers on the cards they drew.

   a. Sample: Some players were close to +1,000,000 but would draw a negative or face card that would prevent them from winning.

   b. Sample: Yes. Positive times positive equals positive. Negative times negative equals positive. Positive times negative equals negative.

2. Choice B

3. Sample: Negative. There is an odd number of negatives.

Graphing Functions

TI-73/TI-83+ Functions

 and **TABLE**

Materials

- Teacher-provided materials: TI-83+ or TI-73 graphing calculators

Demonstration Tips

- This demonstration can be used in addition to or in place of example 3 in Lesson 47.

- Explain to students that the **variable** function represents the input variable and that Y_1 represents the output variable.

- You may want to explore the **TABLE** function in step 1. Press ⌃ repeatedly to see the pairs of values with negative input values. Discuss why the negative input values on the table are not real-world values. **The x-value represents price, and there would not be negative prices at an online auction.**

  Discuss why it would be helpful to get more positive input values from the table. **The selling price might be more than a few dollars, and it would be helpful to know the service charge at the larger selling price.**

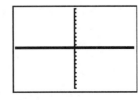

- Students often want to press **GRAPH** to graph the function. Although the graph may be correct, this function key is not always reliable. Instead students should press **ZOOM** and select `ZStandard` when they generate a graph.

- Some students may receive an error display when attempting to graph because the **Plot** function is on. In this case, tell students to select `Quit`, press **2nd** **Y=**, select `PlotsOff`, and then press **ENTER**. The screen should display **Done**.

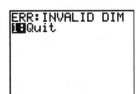

- Students who forget to enter the **variable** into the function may get an incorrect table, a blank graph, or an error display when attempting to graph. Instruct them to select `Quit`, press **Y=**, and reenter the function correctly.

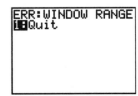

Saxon Math Course 3

- After using the **TRACE** function in step 3, students may need to press ⓐ or ⓑ so that the screen displays **X=0 Y=.3**.

Practice Tips

- For problem 1, students may need help generating the function for their mowing business. Tell them to use the demonstration as a guide.

- Encourage students to use the demonstration steps to find the answers to problem 2.

- In problem 3, discuss which parts of the table or graph would make it unrealistic in the real world. **Sample: There are no negative points in the real world.**

Answers

Step 1:

p	c
1	0.36
2	0.42
3	0.48
4	0.54
5	0.60

Step 2:

Step 3: The relationship is not proportional. The graph does not go through the origin.

1. Check students' work for accuracy.

   a. Sample: We used r for the hourly rate and P for the price we would charge.

   b. Sample: Yes, because our flat fee and hourly rate are less than those of the competition.

   c. Sample: We do not know the cost of mowing, and what we charge the customer may not cover that expense.

2. a.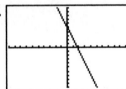

b.

    c. Sample: No, because the graph does not intercept the origin.

3. a. Choice D

    b. Sample: Yes, because the ratio is constant.

    c. Sample: Yes, because a pint is two cups so a cup is 0.5 pints.

Transformations

TI-73/TI-83+ Function

Materials

- Teacher-provided materials: TI-83+ or TI-73 graphing calculators

Demonstration Tips

- This demonstration can be used in addition to or in place of problem 6 in Investigation 5.

- If **L1** and **L2** are not clear before the students are entering the data points, you can instruct them to first go to the **HOME** screen by pressing 2nd and MODE.

    TI-83+ Then have the students press 2nd +, select $\boxed{\text{ClrAllLists}}$ and then press ENTER. The screen should display **Done** .

    TI-73 Then have the students press 2nd 0, select $\boxed{\text{ClrAllLists}}$ and then press ENTER and the screen should display **Done**.

- In step 1, explain why the coordinates of the first point in the list are entered in the list again at the end. **So that the xyLine can complete the connection between all the vertices.**

    If students forget to do this, they will have an incomplete graph. Instruct them to return to the **STAT** menu and reenter the coordinates of the first point again at the end of the list.

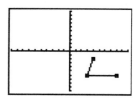

- Discuss why in step 2 we add 7 to **L1** and we subtract 8 from **L2** to get the coordinates of the translated image. **Because moving right adds to the x-values and moving down subtracts from the y-values.**

- Students often want to press GRAPH to graph the translation in step 3. Although the resulting graph may be correct, this function is not always reliable. Instead, students should press ZOOM and select $\boxed{\text{ZStandard}}$ when they generate a graph.

- If the figure and its image seem distorted, tell the tudents to press ZOOM and select $\boxed{\text{ZSquare}}$.

- If other objects show up on the graph, check to see that nothing is entered in the **Y= Editor**. Press Y= and then CLEAR.

- If students receive an error message, it is often due to incorrect entry in the **LIST** function. Tell students to correctly reenter the coordinates by selecting $\boxed{\text{Quit}}$ and going to the **STAT** function.

    TI-83+ Press STAT and select $\boxed{\text{EDIT...}}$.

    TI-73 Press LIST.

- You can demonstrate reflection and rotation with the following formulas. Be aware that once you have entered the original coordinates into **L1** and **L2** and have set the **Plot1** and **Plot2**, you only need to calculate the coordinates of the translation and graph other transformations.

Reflection over the *x*-axis

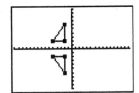

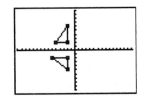

Reflection over the *y*-axis

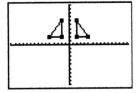

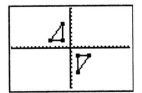

Rotation of 90 degrees

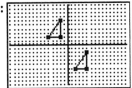

Rotation of 180 degrees

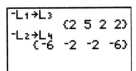

Practice Tips

- Students may want to have the same display as the graph for problem 2. Tell them to press **2nd** **ZOOM**, ⌄ once and ⌃ once, then **ENTER**.

- Tell students that they first need to find how much the figure is translated left or right and up or down. Review the quadrants of the coordinate plane to help determine to where the rectangle is moving. Then calculate the distance that the furthest point on each object is from the origin to calculate how far the figure is translated.

Answers

Step 2: L1 + 7 ⟶ L3; L2 – 8 ⟶ L4

Step 4:

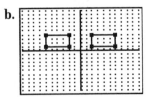

1. **a.** L1 + 5 ⟶ L3; L2 + 6 ⟶ L4

   b.

   c. (–3, 0)

2. **a.** L1 – 8 ⟶ L3; L2 ⟶ L4

   b.

   c. reflection across the *y*-axis

Saxon Math Course 3

Scientific Notation for Small Numbers

TI-73/TI-83+ Functions

$\boxed{\text{Sci}}$, $\boxed{\wedge}$, and $\boxed{(-)}$

Materials

- Teacher-provided materials: TI-83+ or TI-73 graphing calculators, adding machine tape

Demonstration Tips

- You may choose to use the demonstration to reinforce the solutions to examples 6 and 7 in Lesson 51 instead of replacing them.

- Remind students that $\boxed{\wedge}$ denotes **power.**

- Some students may discover that they do not have to enter $\boxed{0}$ before the decimal or after the ones place. Encourage them to practice the proper form.

- Be aware that the calculator will not convert numbers in scientific notation with a power smaller than −3 to standard form. The screen will display a number with the **E** function in **Normal** or **Scientific Mode.**

```
1.0*10^-4
            1E-4
```

- For step 1, you can have the students use the different keystrokes.

  TI-83+ Press $\boxed{1}$ $\boxed{\cdot}$ $\boxed{5}$ $\boxed{\text{2nd}}$ $\boxed{,}$ $\boxed{(-)}$ $\boxed{3}$ and $\boxed{\text{ENTER}}$.

  TI-73 Press $\boxed{1}$ $\boxed{\cdot}$ $\boxed{5}$ $\boxed{\text{2nd}}$ $\boxed{\wedge}$ $\boxed{(-)}$ $\boxed{3}$ and $\boxed{\text{ENTER}}$.

```
1.5E-3
            .0015
```

- Discuss what it means for a number to have a negative exponent. **The number is divided by 10 to that power.**

- For step 2, if the students wish to confirm that **E** symbolizes ×10 in scientific notation, have them use the form in the calculator by using 7.0, $\boxed{\times}$, 10, $\boxed{\wedge}$, $\boxed{(-)}$, and 6. Then press $\boxed{\text{ENTER}}$.

```
7.0*10^-6
            7E-6
```

Practice Tips

- For problem 1, you may want to use the board instead of adding machine tape. Call out the numbers listed to the right in order and one at a time. Remind students to switch from **Normal** to **Scientific** modes, as needed.

8.6×10^{-1}
5.42×10^{-4}
1.024×10^{-8}
0.00157
0.0024
0.000006987
3.14×10^{-2}
0.0045

- In problem 2, some students may not understand the meaning of "bit rate". Explain to them that a byte is the smallest form of information in a computer. The amount of time in seconds it takes a computer to process one byte of information is called the bit rate. A bit rate of one nanosecond means that a computer processes one byte of information in 0.000000001 second.

- Be aware that in problem 3 students may forget to multiply 1.0×10^{-6} by 2.5.

Answers

Step 1: 0.0015

Step 2: 7.0×10^{-6}

1. Check students' work for accurate positioning on the number line.

   a. Sample: To place the numbers, we needed to look at the power of the 10. We also had to remember that the more negative an exponent meant the number was closer to 0.

   b. Sample: 1.024×10^{-8} and 0.000006987 because the other numbers were already too close to 0 and these two needed to be closer.

   c. Sample: All of the numbers are the same except today's numbers are in decimal form.

2. Choice A

3. 2.5×10^{-6} m; 0.0000025 m

4. Yes, because the smaller the exponent, the smaller the number.

Collect, Display, and Interpret Data

TI-73 Functions

▯▯▯, ⊘, and ▯▯▯

Materials

• Teacher-provided materials: TI-83+ or TI-73 graphing calculators

Demonstration Tips:

• This demonstration can be used in addition to or in place of problem 7 in Investigation 6.

• This demonstration is for the **TI-73** graphing calculator. The TI-83+ is only equipped with the **histogram** function.

• In step 1, if **L1** and **L2** are not clear before the students enter the data points, instruct them to first go to the **HOME** screen by pressing (2nd) and (MODE). Then have them press (2nd) (0), select ⎵ClrAllLists⎵, and press (ENTER). The screen should display **Done**.

• For step 2, students using the TI-83+ can generate a bar graph using the **histogram** function. Instruct them to follow step 2, which is similar to the steps needed for the TI-83+. However, make them aware that the histogram is not a true bar graph.

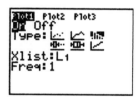

• At step 3, students may want to use the **ZOOM** function instead of (GRAPH). Explain that the calculator will automatically zoom for these types of graphs. They should still use **ZOOM** for other types of graphs.

• If other objects appear on the graph, check that nothing is entered in the **Y= Editor**. Press (Y=) and then (CLEAR).

• If students receive an error message due to an incorrect entry in the **LIST** function, tell them to correctly reenter the coordinates after selecting ⎵Quit⎵ and pressing (LIST).

• Although students use the bar graph as their answer to problem 7, you may encourage them to complete all graphs in step 2 for practice. Below are the graphs as they should appear on the screen.

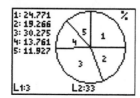
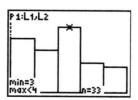

- If you do not practice all of step 2, tell students to refer to it for the practice problems.
- Explain that the values in the first column/row in any table are usually entered into **L1** and that the values in the second column/row are entered into **L2**.

Employee	Total Pay
Jan	$230
Stephen	$440
Michael	$185
Rosie	$385

L1	L2	L3	2
1	230	------	
2	440		
3	185		
4	385		

L2(5) =

Day	MON	TUE	WED	THR	FRI
Speeding Tickets	120	85	43	51	97

L1	L2	L3	1
1	120	------	
2	85		
3	43		
4	51		
5	97		

L1(6) =

- Students will need to set the **ZoomStat** function for problem 3. Tell them to press ⎡Zoom⎤ and select ⎡ZoomStat⎤.

Answers

Steps 1–3: Bar graph; Bar graphs and circle graphs are appropriate for qualitative (or categorical) data. However, sectors of circle graphs represent fractions of the whole, and in this case, students were allowed to select more than one option in the survey. Therefore, a circle graph would not be an appropriate fit for the data.

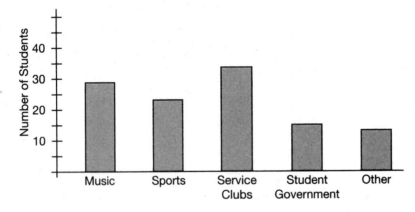

1. Choice B

2. a.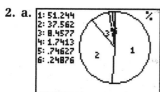

b. Sample: Trucks and cars, because they make up over 88% of the data.

c. Sample: The sample is from only those students who are driven to school. A sample from all students who walk, ride the bus, and are driven by their parents would be more representative.

3. a. Circle graph; Sample: A circle graph only shows the percentages.

b.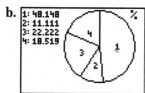

c. Since the actual amount of money spent for each type of expense is now shown in the circle graph, an advisor would not have enough information from the graph to offer knowledgeable advice.

Direct Variation

TI-73/TI-83+ Functions

 **Y=**, **TABLE**, and **WINDOW**

Materials

• Teacher-provided materials: TI-83+ or TI-73 graphing calculators

Demonstration Tips

• This demonstration can be used in addition to or in place of example 1 in Lesson 69.

• For step 2, students who forget to enter the **variable** into the function may get an incorrect table. Instruct those students to select Quit, press **Y=** and then reenter the function correctly.

• At step 3, students may use the **ZOOM** function instead of **GRAPH**. Explain that the calculator will automatically zoom for this type of graph. However, it is still recommended to use **ZOOM** for other types of graphs.

• Some students may receive an error when attempting to graph because the **Plot** function is on. Tell them to select Quit, press **2nd** **Y=**, select PlotsOff, and press **ENTER**. The screen should display **Done**.

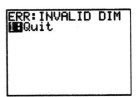

• Review the terminology for domain and range. Discuss which values are appropriate for most real-world scenarios. **The x- and y-values in the first quadrant.**

Discuss why the negative input values on the table are not real-world values. **The x-value represents miles, and there would not be a negative amount miles for a taxi ride.**

• Instruct students to go through steps 1–3 carefully to generate the tables and graphs for the remaining functions. All of the steps should be the same except for the specific values and functions that need to be entered. Remind students to use **x²** or **^** **2** for the power of 2.

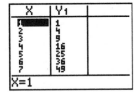 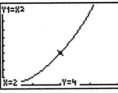

Functions for Scenario B

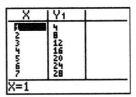 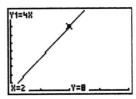

Functions for Scenario C

Practice Tips

• Encourage students to follow the steps from the demonstration for problem 1. Discuss what values for x and y should be used in the **WINDOW** function. Sample: 0 to 30 for x; 0 to 200 for y.

- In problem 2c, show students how to use the **Table Set** function to find large values of x or y. Press .

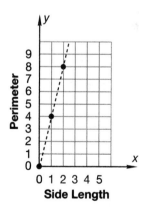

- Encourage students to use the functions in the graphing calculator to solve problem 3. Be aware that the problem is asking students to find the x-value of a given y-value.

Answers

Step 2:

$D = 2m + 3$

Miles	Dollars	$\frac{\text{Dollars}}{\text{mile}}$
1	5	$\frac{5}{1}$
2	7	$\frac{7}{2}$
3	9	$\frac{9}{3}$
4	11	$\frac{11}{4}$

$A = s^2$

s	A	$\frac{A}{s}$
1	1	$\frac{1}{1}$
2	4	$\frac{4}{2}$
3	9	$\frac{9}{3}$
4	16	$\frac{16}{4}$

$P = 4s$

s	P	$\frac{P}{s}$
1	4	$\frac{4}{1}$
2	8	$\frac{8}{2}$
3	12	$\frac{12}{3}$
4	16	$\frac{16}{4}$

Step 3:

A Taxi

$D = 2m + 3$

B Area

$A = s^2$

C Perimeter

$P = 4s$

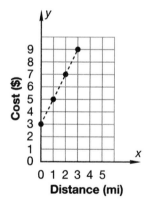

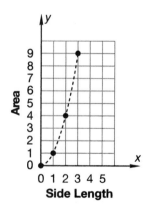

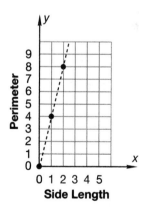

Only a graph in which the points are aligned (linear), rise to the right, and align with the origin depicts a proportional relationship. Graph **B** is not linear. Graphs **A** and **C** are linear, but graph **A** does not align with the origin.

1. a. $b = 3.50m + 43.95$

　　b. Check for students' work on constant ratio.

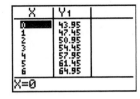

　　c. Check for students' complete labeling.

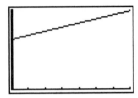

　　d. No, because it does not align with the origin.

2. a. Check for students' work on constant ratio.

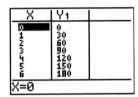

　　b. Yes, because the ratio is constant.

　　c. Choice B

3. Choice A

Saxon Math Course 3

Probability Simulation

TI-73/TI-83+ Function

 MATH randInt(

Materials

- Teacher-provided materials: TI-83+ or TI-73 graphing calculators

Demonstration Tips

- This demonstration can be used in addition to or in place of the second activity in Investigation 7.

- For step 1, make sure that the students assign a number to each possible outcome. For instance, if they simulate the basketball player's free throws, the students could assign {1, 2, 3, 4} to making baskets and {5, 6} to missing.

- Students may misunderstand that they have to use the integers from the demonstration. Tell them that they can choose any sequential integers, even negative ones, in their simulation.

- In step 2, students may discover that they do not need ▬ for the **random integer** function. Encourage them to use the proper form.

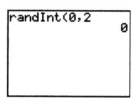

- At step 3, instruct students to record the results from the calculator screen as they conduct the trials, especially if they do more than 6.

- Advise students to use the chart from the first activity as a template for the chart they must generate for this activity.

- Some students may have identical random numbers, especially if the calculator memory has been reset. Tell the students to press ENTER several times in the **HOME** screen. Then reset the **random integer** function for their simulation.

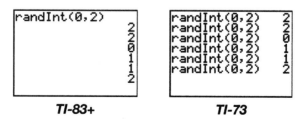

| TI-83+ | TI-73 |

Practice Tips

- For problem 1, you may need to explain "raffle". **A raffle is an event where people receive a numbered ticket and an identical ticket is placed in a box, spinner, or small barrel. Then one ticket is drawn, and the person holding the matching ticket wins a prize.**

- Review the concept of odds. **The first number represents the number of favorable chances, and the second number represents the number of unfavorable chances. The total number of possibilities is the number of favorable plus unfavorable outcomes.**

• To answer problem 2c, students may need help seeing that the second event is simply reduced numbers from the first event.

Answers

Step 1: Samples: rolling 1–4 he makes baskets, rolling 5–6 he misses; 1 and 2 for blue socks, 3, 4, 5 and 6 for black socks; 1 represents a winning seat, 2 a non-winning seat.

Step 3: Check students' work for results chart and experimental probability calculation.

1. **a.** Sample: 1 is Danny's ticket; 2–1000 are other tickets.

   b. Check students' charts based on their simulation settings.

   c. Sample: If each ticket had a price, it would be expensive to simulate a raffle drawing.

2. **a.** Sample: 1–4 are the boys; 5–14 are the girls.

   b. Sample: 1 and 2 are the boys; 3–7 are the girls.

   c. No, because the experiment in b. is the same as the one in a., except that it uses reduced numbers.

Formulas for Sequences

TI-73/TI-83+ Function

`seq(`

Materials

- Teacher-provided materials: TI-83+ or TI-73 graphing calculators

Demonstration Tips

- This demonstration can be used in addition to or in place of example 1 in Lesson 73.

- Some students may confuse **xth** and **nth**. Discuss how mathematical notation may be different on the calculator than in the book.

- Students may discover that they do not need 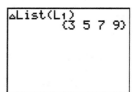 for the **sequence** function. Encourage them to use the proper form.

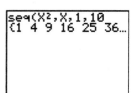

- The **LIST** function can calculate the common difference between the terms. If the difference is not constant but there is still a pattern, then there may still be a rule that is a combination of operations.

  TI-83+ After finding the terms with the **sequence** function, press `STO▸` `2nd` `1` then `ENTER`. Press `2nd` `STAT` and `▶` once then select `List(`. Press `2nd` `1` `)` then `ENTER`.

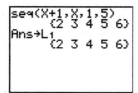 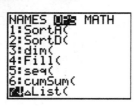 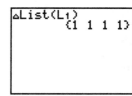

  TI-73 After finding the terms with the **sequence** function, press `STO▸` `2nd` `LIST` `1` then `ENTER`. Press `2nd` `LIST` and `▶` once then select `List(`. Press `2nd` `LIST` `1` `)` then `ENTER`.

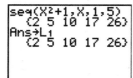 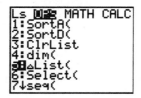 

Practice Tips

- For problem 1, when entering the equation into the **sequence** function, students may get confused by the extra parentheses. Help them understand that the outer parentheses are for the function in the graphing calculator and that the inner parentheses are for the expression $n(n + 1)$.

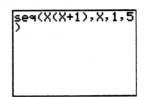

- To answer problem 2a. with the graphing calculator, tell students to use the **sequence** function to find the first five terms of each choice.

- To answer problem 2b., students will need to set the range of terms in the **sequence** function to 100 and 100.

- Students may receive an error message if they forget to enter the variable or the position number. In either case, instruct students to select Quit and to reenter the sequence.

seq(X²,X,100)■	ERR:ARGUMENT ┃1█Quit 2:Goto	seq(X²,100,100)	ERR:SYNTAX ┃1█Quit 2:Goto

TI-83+   TI-73

- The fractions in the sequences may give some students difficulty for problem 3. Remind them that they should use parentheses around the numerator and denominator of a fraction. Also remind students to use ⌃ to denote power.

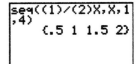

Answers

Step 1: {3, 6, 9, 12, 15, 18, 21}

Step 2: {3, 6, 9, 12, 15, 18, 21, 24, 27}

Step 3: {24, 27}

1. **a.** seq(x (x + 1), x, 1, 5)

   b. {2, 6, 12, 20, 30}

   c. {132, 156, 182, 210, 240}

2. **a.** Choice D

   b. 10,200

3. Choice B

Scatterplots

TI-73/TI-83+ Functions

⌐∵⌐ , ⌐∠⌐ , and $\boxed{\text{LinReg(ax+b)}}$

Materials

- Teacher-provided materials: TI-83+ or TI-73 graphing calculators, stop watches

Demonstration Tips

- This demonstration can be used in addition to or in place of examples 2 and 3 in Investigation 8.

- If **L1** and **L2** are not clear before the students enter the data points in step 1, instruct them to first go to the **HOME** screen by pressing ⬤ₙd and ⬤ₘₒₐₑ.

  TI-83+ Then have the students press ⬤ₙd ⬤₊, select $\boxed{\text{ClrAllLists}}$, and press ⬤ₑₙₜₑᵣ. The screen should display **Done**.

  TI-73 Then have the students press ⬤ₙd ⬤₀, select $\boxed{\text{ClrAllLists}}$, and press ⬤ₑₙₜₑᵣ. The screen should display **Done**.

- When setting the calculator to plot the pairs of numbers on the coordinate plane in step 2, discuss the connection between **L1** and the *x*-coordinates and between **L2** and the *y*-coordinates. You may also want to discuss that the values in the first column/row in any table are usually entered into **L1** and that the values in the second column/row are entered into **L2**.

- Students may want to press ⬤ₐᵣₐₚₕ to plot the pairs of numbers in step 3. Although the resulting graph may be correct, this function key is not always reliable. Instead, students should press ⬤ᵤₒₒₘ and select $\boxed{\text{ZoomStat}}$ every time the students graph.

- To ensure the accuracy of their graphs, students should use the calculator graph as a guide and the table for the pairs to plot.

- At step 4, some students may make the error of sorting only one of the lists or sorting both lists separately. These students will need to reenter the data in the **STAT** function so that the pairs of data will remain ordered pairs.

- In step 7, some students using the TI-83+ may select 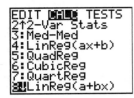. These students will need to redo the **regression** function so that $\boxed{\texttt{LinReg(ax+b)}}$ is selected.

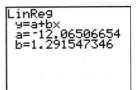

- When the calculator finds a regression equation (best-fit line) with a negative *y*-intercept, discuss how it relates to the real world. **Sometimes the data is scattered enough that the best-fit line will have unrealistic values. Because the line fits the overall behavior of the data, we understand that there are limits to the values that the regression can represent.** For instance, based on the best-fit line, if you make –$12 with a high school diploma, someone with a college degree makes $0.

- To enter the regression equation in step 8, some students may forget to use ⊞ and ⟨(−)⟩ and will receive an error message. Instruct those students to select $\boxed{\texttt{Quit}}$ and then reenter the equation correctly.

- Students who forget to enter the **variable** into the function may get a scatterplot with no line graph. Instruct those students to select ⟨Y=⟩ and then reenter the equation correctly.

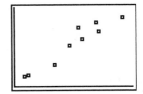

Practice Tips

- In problem 1, round the seconds to the nearest hundredth of a second. Discuss how a hundredth of a second relates to seconds and minutes. **Each hundredth to a second is like a second to a minute, except it takes 100 hundredths to equal 1 second.**

- Discuss how data collected on the "wave" could be used in the real world. **Sample: It could be used to make the wave of an audience look realistic in computer generated special effects for a football movie.**

- Discuss what kind of correlation is displayed in the scatterplot for problem 2: positive or negative. **Negative.** Ask students to explain how this may be true for the age group of the data. **Sample: As a baby you need a lot of sleep, but as you get older, you need less sleep.**

 Saxon Math Course 3

Answers

Steps 1–3:

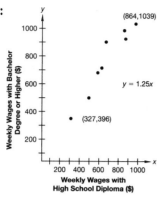

Steps 4–6: Yes, they are correlated because most of the connected lines are increasing.

Step 7: $y = 1.29x + (-12.1)$

Step 8:

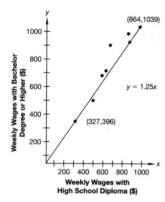

1. **a.** Check students' work.

   b. Yes, because the overall behavior of the data shows a positive correlation.

   c. Check students' work.

2. **a.** Yes, the data indicate a negative correlation.

   b. $y = -0.53x + 15.0$

   c.

   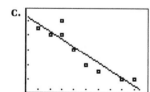

Graphing Equations Using Intercepts

TI-73/TI-83+ Functions

STAT and Y=

Materials

- Teacher-provided materials: TI-83+ or TI-73 graphing calculators

Demonstration Tips

- This demonstration can be used in addition to or in place of example 1 in Lesson 82.

- If **L1** and **L2** are not clear before the students enter the data points in step 1, you can instruct them to first go to the **HOME** screen by pressing 2nd and MODE.

  TI-83+ Then have the students press 2nd +, select ClrAllLists, and press ENTER. The screen should display **Done**.

  TI-73 Then have the students press 2nd 0, select ClrAllLists, and press ENTER. The screen should display **Done**.

- If students have difficulty transforming the equation in step 2, refer them to Lesson 79, Transforming Formulas.

- Be aware that students may forget to divide every term by 4 when transforming to slope-intercept form.

$$3x + 4y = 24$$
$$4y = -3x + 24$$
$$y = -\frac{3}{4}x + 24$$

- Students may realize that they do not have to use the parentheses for the fraction. Encourage them to use the parentheses to separate the fraction from the variable.

- Students may want to press GRAPH to plot the pairs of numbers. Although the resulting graph may be correct, this function key is not always reliable. Instead, students should press ZOOM and select ZStandard.

- Students who forget to enter the **variable** into the equation may get an error message. Instruct them to select Quit, press Y=, and then reenter the equation correctly.

Practice Tips

- Be sure that the students use the space provided to work their calculations for problems 1–2.

- Some students may forget to divide by the negative when transforming the equation $2x - 5y = 30$ to the slope-intercept form in problem 1.

$$-5y = -2x + 30$$
$$y = -\frac{2}{5}x + 6$$

- If the graph does not fully fit the display using **ZStandard** have the students press ZOOM and ZSquare.

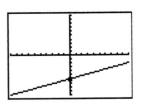

Saxon Math Course 3

- Discuss what it means when the *y*-intercept is (0, 30) as in $y = 1.5x + 30$ from problem 2. It can mean that the beginning value is 30 for a real-world application.

Answers

Example 1: The *y*-intercept is (0, 6); the *x*-intercept is (8, 0).

Step 2: $y = -\frac{3}{4}x + 6$

Step 3: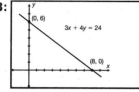

1. **a.** The *y*-intercept is (0, −6), the *x*-intercept is (15, 0).

   b. $y = \frac{2}{5}x - 6$

   c.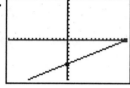

2. **a.** The *y*-intercept is (0, 30); the *x*-intercept is (−20, 0).

   b. $y = 1.5x + 30$

   c.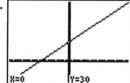

Solving Problems with Two Unknowns by Graphing

TI-73/TI-83+ Functions

 Y= , TRACE , and TABLE

Materials

- Teacher-provided materials: TI-83+ or TI-73 graphing calculators

Demonstration Tips

- This demonstration can be used in addition to or in place of the example in Lesson 89.

- If students have difficulty transforming the equation in step 2, refer them to Lesson 79, Transforming Formulas.

- Students may press GRAPH to graph the system. Although the resulting graph may be correct, this function key is not always reliable. Instead, students should press ZOOM and select $\boxed{\text{ZStandard}}$ to graph.

- Students who forget to enter the **variable** into the function may get an error message or erroneous table. Instruct those students to select $\boxed{\text{Quit}}$, press Y= , and then reenter the function correctly.

- There may be an error message if students use − instead of $\boxed{(-)}$. Tell them to select $\boxed{\text{Quit}}$ and press Y= to correctly reenter the equation.

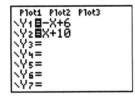
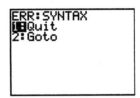

- In step 3, explain that the graph on the calculator is an approximation of the line equations and that the table gives the exact values for each line.

- Explain why **Y1** and **Y2** are the same value. **When two equations share the same point, then that point has the same x- and y-values.**

Practice Tips

- Be sure that the students show their work in the space provided for problems 1–2.

- In problem 2, the graph may not show up with ZStandard. Tell students to press ZOOM and select $\boxed{\text{ZoomFit}}$. The graph may still not show up, at least with the point of intersection. Tell students to press ZOOM , select $\boxed{\text{Zoom Out}}$, then press ENTER .

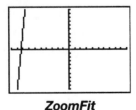
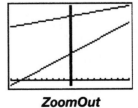

ZoomFit   *ZoomOut*

Answers

Example: $y = -x + 6$; $y = x + 10$

Step 2:

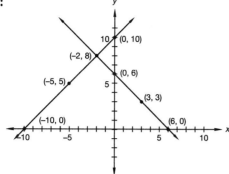

Step 3: $(-2, 8)$

1. a. $y = -2x - 2$; $y = 2x + 6$

   b.

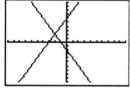

   c. $(-2, 2)$

2. a. $y = 15x + 125$; $y = 5x + 305$

   b.

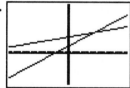

   c. $(18, 395)$; The band must sell 18 tickets to cover the cost of $395.

Inverse Variation

TI-73/TI-83+ Functions

Y= and **TABLE**

Materials

- Teacher-provided materials: TI-83+ or TI-73 graphing calculators

Demonstration Tips

- This demonstration can be used in addition to or in place of example 2 in Lesson 99.
- If students have difficulty transforming the equation in step 2, refer them to Lesson 79, Transforming Formulas.
- In step 1, if there are other functions entered in the **Y= Editor**, students can press **CLEAR**.
- Explain that the **variable** function represents any input variable and that Y_1 represents any output variable.
- For step 2, students who forget to enter the **variable** into the function may get an error message. Instruct them to select Quit, press **Y=**, and reenter the function correctly.

- Students may have trouble finding pairs of values in whole numbers. Tell them that the table should have *x*- and *y*-values that are not decimals.
- At step 3, students may want to use the **ZOOM** function instead of **GRAPH**. Explain to students that the calculator will automatically zoom for this type of graph but that they should use **ZOOM** for other types of graphs.
- Discuss which values are appropriate for most real-world scenarios. **The x and y-values in the first quadrant.**

  If you use ⬆ in the **TABLE** function, discuss why **ERROR** is in Y_1 when **X** is **0**. **Because $\frac{80}{0}$ is undefined.**

  You can also look at the *y*-values for the negative *x*-values. Discuss why these are not appropriate for the real-world scenario in the example. **The *x*-value (*n*) represents the number of robots which would not be a negative number.**

Practice Tips

- In problem 1, if students are having difficulty setting the **WINDOW** for the graph, suggest using 0 for **Xmin**, 7 for **Xmax**, 0 for **Ymin**, and 35 for **Ymax**.
- Students may need help setting up the table in problem 2. Use the answer key to give them the *b*-values of the table.
- Use ⬆ and ⬇ in the **TABLE** function to find the values for *d*.

Saxon Math Course 3

Answers

Example: $n \cdot t = 80$; $t = \frac{80}{n}$

Step 2:

n	1	2	4	5	8	10	16	20	40	80
t	80	40	20	16	10	8	5	4	2	1

Step 3: ; 4 hours

1. a. $s \cdot t = 60$

  b. $t = \frac{60}{s}$

  c.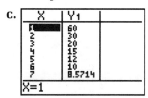

  d.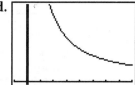

  e. 12 hours

2. a. $d = \frac{56}{b}$

  b.

  c. Sample: No, because for another half hour of break time the employee improves only slightly.

Box-and-Whisker Plots

TI-73/TI-83+ Functions

 and [TRACE]

Materials

- Teacher-provided materials: TI-83+ or TI-73 graphing calculators, number cubes

Demonstration Tips

- This demonstration can be used in addition to or in place of examples 1 and 2 in Lesson 103.

- If **L1** is not clear before students enter the data points, instruct them to first go to the **HOME** screen by pressing [2nd] and [MODE].

  TI-83+ Then have the students press [2nd] [+], select [ClrAllLists], and press [ENTER]. The screen should display **Done**.

  TI-73 Then have the students press [2nd] [0], select [ClrAllLists], and press [ENTER]. The screen should display **Done**.

- Students may want to press [GRAPH] to plot the pairs of numbers in step 3. Although the resulting graph may be correct, this function key is not always reliable. Instead, students should press [ZOOM] and select [ZoomStat] for graphs from the **PLOT** function.

- If other objects show up on the graph, check to see that nothing is entered in the **Y= Editor**. Press [Y=] and then [CLEAR].

Practice Tips

- In problem 1, students should find the medians of the first and second halves by covering up the second half of the box-and-whisker plot and finding the middle line. **Quartile 1 is the median of the first half.** Then have the students cover up the first half of the plot and find the middle line. **Quartile 3 is the median of the second half.**

- Explain that the extremes are sometimes called "outliers" because they are far from the majority of the data. Ask how an outlier affects the box-and-whisker plot. **It makes one whisker longer than the other.**

- Discuss why information like this may be important. **If you are raising chickens, you can keep track of the weights and look out for sudden changes, which will help stay on top of illness or over-population.**

Answers

Steps 1–3:

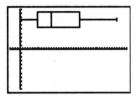

1. **a.** Check students' work.

   b. Check students' work to verify that the median is the same as the middle mark of the plot.

   c. and d. Check students' work to verify that the median of the first half is the same as the first quartile mark and that the median of the second half is the same as the third quartile mark.

2. **a.** Check students' work for labels: minX=0, Q1=7, Med=13, Q3=24.5, and MaxX=40.

   b. 13 hours; The median is a measure of central tendency, which means that most students tend to study around 13 hours a week.

   c. 0 and 40; The 40 is less like the rest of the data because it is the farthest away from the median.

3. Choice A

Consumer Interest

TI-83+ Function

`TVM Solver...`

Materials

- Teacher-provided materials: TI-83+ or TI-73 graphing calculators

Demonstration Tips

- This demonstration can be used in addition to or in place of the example in Lesson 109.

- This demonstration is for the **TI-83+** graphing calculator. The TI-73 is not equipped with the **TVM Solver** function.

- Explain what each part in the **TVM Solver** means. **N is the number of months, I% is the annual interest rate, PV is the amount borrowed, PMT is the monthly payment, and FV is the remaining balance. P/Y stands for payments per year, and C/Y stands for compounds per year. The PMT: END or BEGIN indicates when interest is calculated and payments are made in the month.**

- In step 1, be aware that some students may incorrectly convert the interest rate to a decimal for **I%=**. They will not receive an error message but an incorrect **FV**. In addition, the interest rate is 24 because 2% is a monthly rate and **I%** is an annual rate, $2 \times 12 = 24$.

- When a function is solved in the **TVM Solver**, a ■ icon will appear.

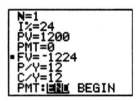

- The **FV** is negative to indicate cash flow. After one month, Jason still owes money on his credit card, which is a negative cash flow.

  The **PV** is positive to indicate cash flow. At the beginning of the month, the credit card lends Jason money to make the purchases he wants, which is a positive cash flow.

- To find the amount of interest, tell students to use FV − PV.

- For step 2, the **PMT** is negative to indicate cash flow. At the end of the month, Jason makes a payment of $50, which is a negative cash flow.

- Students may use ⬛➖ instead of `(−)` for negative cash flows. This will result in an error message. Tell them to select `Quit`, press **APPS**, and select `Finance` and `TVM Solver...`.

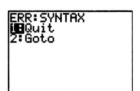

- At step 3, students must find the range that the remaining balance could be, depending on when the additional purchases were made. If they were made at the end of the month, then the interest calculation would be minimal, almost as little as if no purchases were made. If the purchases were made at the beginning of the month, then the interest calculation would be at a maximum, calculated off the previous balance and the additional purchases. Thus, the two calculations with **TVM Solver** find the lowest to the highest balance Jason could have for the next month.

Practice Tips

- In problem 1, be sure that the students annualize the monthly interest rate, 2.499 × 12 = 29.988.

- When the students solve for the **FV**, have them write down exactly what the calculator displays including the negative sign.

- The students should drop the negative sign on the **FV** and round it to the nearest hundredth in order to calculate the amount of interest.

- For problem 2, be sure that the students annualize not only the **I%** but the **N**, 1 × 12 = 12.

- For problem 3, be sure that the students use the given **I%** but annualize the **N**, 4 × 12 = 48.

- Students will need to add the cost of tuition and books to get the **PV**.

Answers

Step 1: $24

Step 2: $1174

Step 3: from $1997.48 to $2013.48

1. a.

   b. −5534.946

   c. $134.95

2. a.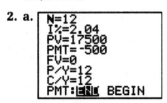

   b. −11803.93782

   c. Choice C

3. a.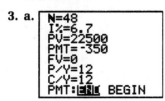

   b. −10187.92025

   c. $10,187.92

   d. Sample: Make a larger payment.

Non-Linear Functions

TI-73/TI-83+ Functions

 Y= , WINDOW , and TRACE

Materials

- Teacher-provided materials: TI-83+ or TI-73 graphing calculators

Demonstration Tips

- This demonstration can be used in addition to or in place of the second activity in **Investigation 11**.

- Students who use ⊟ instead of (−) in step 1 may receive an error message. Tell those students to select Quit, press Y= , and reenter the function correctly.

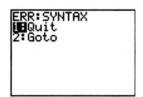

- You may wish to explore the **TABLE** function. Press ⌒ repeatedly to see the pairs of values with negative input values. Discuss why the negative input values on the table are not real-world values. **The x-value represents time, and time cannot be negative.**

- At step 2, students may want to use the **ZOOM** function instead of GRAPH Explain that **WINDOW** zooms to the area on the coordinate plane for this demonstration, but that students should still use **ZOOM** for other types of graphs.

- Discuss why gravity has a negative value. **Because gravity pulls on objects and reduces an object's speed.**

- Students who forget to use X,T,θ,n with x^2 will get an incorrect table and no graph. Instruct those students to press Y= and reenter the function correctly.

 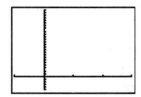

- If the students receive an error message or plots on a graph, instruct them to turns the plots off by pressing 2nd Y= , selecting PlotsOff, and pressing ENTER.

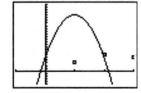

Practice Tips

- For problem 1, discuss the coefficients in the function. **−16 is gravitational pull, 125 is speed, and 2 is the height of the object from the ground.**

- To set **WINDOW**, suggest x-values from −1 to 7 and y-values from −25 to 175.

Saxon Math Course 3

- Explain that the time for when the ball hits the ground is just before the y-value is negative.
- In problem 2, ask students if they have ever been to a concert and have them discuss their observations regarding attendance and time of day.
- Suggest that students find the maximum and minimum values by using the table only.
- Remind students that the x-values are added to 6 p.m. to find the time of day of the concert.
- For problem 3, connect the discussion from problem 1 on choosing the correct function.
- To set **WINDOW**, suggest x-values from −1 to 5 and y-values from −10 to 70.

Answers

Step 1:

t	h
0	5
1	21
2	5

Step 2:

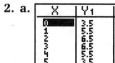

Step 3: 21 feet; 1 second; 2.14 seconds

1. a.

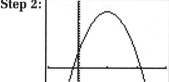

b.

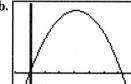

c. 158 feet

d. 6.3 seconds

2. a.

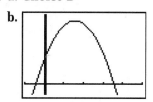

b. between 8 p.m. and 9 p.m.; Sample: The best bands come on at sunset.

c. after midnight; Sample: Noise ordinances prohibit concerts after midnight.

3. a. Choice B

b.

c. 64 ft; 1.5 seconds

Significant Digits

TI-73/TI-83+ Functions

QuadReg

Materials

- Teacher-provided materials: TI-83+ or TI-73 graphing calculators

Demonstration Tips

- This demonstration can be used in addition to the explanation on significant digits in Lesson 117.

- Students may want to press GRAPH to plot the pairs of numbers in step 3. Although the resulting graph may be correct, this function key is not always reliable. Instead, students should press ZOOM and select ZoomStat every time they plot graphs.

- In step 4, some students using the TI-83+ may select QuartReg. These students will need to redo the **regression** function so that QuadReg is selected.

- Be sure that the students use the correct rounding for significant digits in step 4.

- Students who forget to use X,T,Θ,n with x² will get an incorrect table and graph. Instruct those students to press Y= and reenter the function correctly.

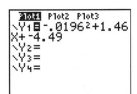 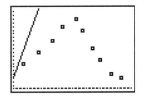

- Students may get confused when + and − are right next each other in the regression equation. Tell those students to use (and) around the negative number to clear up any confusion.

- Remind students that a regression equation is an average of the data points. A best-fit equation should go through the middle of most of the points.

Practice Tips

- Encourage students to follow the demonstration steps for problem 1.

Answers

Step 3: quadratic; The points increase and then decrease.

Step 4: $y = -0.1x^2 + 2.0x + (-2.0)$

Step 5: 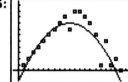 ; No, because it does not go through the middle of the points.

Step 6: 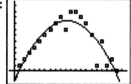 ; Yes, because it goes through the middle of the points.

1. a. quadratic; The points make an upside down U-shape.

   b. $y = -0.020x^2 + 1.5x + (-4.5)$

   c.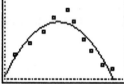

   d. Yes, because the regression goes through the points.

Sine, Cosine, Tangent

TI-73/TI-83+ Functions

 **SIN** , **COS** , and **TAN**

Materials

- Teacher-provided materials: TI-83+ or TI-73 graphing calculators

Demonstration Tips

- This demonstration can be used in addition to or in place of example 3 in Lesson 118.

- In step 1, students should verify that the calculator is in **Degree Mode**, especially for those using a TI-83+.

- Some students may discover that they do not have to use **)** to calculate the sine of the angle. Encourage those students to use the proper form.

- To help with rounding, students can set the **MODE** to three decimal places. Tell them to press **MODE**, ⌄ once, ▶ four times, and then **ENTER**. This function will round all numbers calculated to the thousandths place.

- For step 2, students may be curious about **COS⁻¹**. Tell them that the **inverse trig** functions are used to find an angle with given side lengths.

- For step 3, explain that the angle of elevation is usually on the bottom vertex from the right angle.

- Some students may need help setting up the proportion to solve for the height. Refer them to the Trigonometry Ratios at the beginning of Lesson 118.

Practice Tips

- For problem 1, review special right triangles. Make connections between the similar trig ratios and the similar lengths of a 45-45-90 triangle.

- In problem 2, students may need help explaining why tangent would not be used. Discuss which lengths are given and which lengths are needed to solve the problem. **The opposite side length is given and the hypotenuse length is needed. Tangent is only for the opposite and adjacent side lengths.**

- For problem 3, students may need help deciding on which trig function to use. Ask them which length is given and which length is needed to solve the problem. **The adjacent side length is given and the hypotenuse is needed.**

  Ask the students which trig ratio uses the given and needed lengths. **Cosine is adjacent over hypotenuse.**

Answers

Step 1: 0.574

Step 2: 0.819;

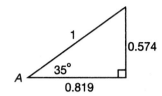

Step 3: 0.404; 40.4 ft

1. a. 0.707

   b. 0.707

   c. 1

   d. They are equal. If one angle is 45° and the other angle is 90°, then the last angle is 45°. These are the angles of an isosceles triangle, which means that two legs are equal.

   e. Tangent is the ratio of the opposite over adjacent lengths. If those lengths are equal then the tangent is a 1 to 1 ratio.

2. a. 0.940

   b. 22 ft

   c. No, because we have only the opposite side length and need the hypotenuse.

3. Choice B

• Measures of Central Tendency

New Calculator Functions: STAT and LIST

Example: Below are the prices of new homes sold in a certain neighborhood. Find the mean, median, and mode of the data. Which best represents the data?

Home Prices (in thousands $)

170	191
208	175
185	175
209	187
181	195
183	219

Prices of New Homes Sold (in thousands $)

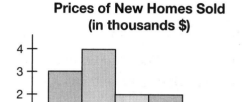

Demonstration

Step 1   To find the mean, median, and mode of the data, first enter the list of numbers into the calculator.

• Go to the **STAT** menu.

TI-83+   Press STAT and select EDIT... .

TI-73   Press LIST .

• Enter the Home Prices from example 5 into **L1** by entering 170 and pressing ENTER , entering 191 and pressing ENTER , and so on. Your screen should display **L1(13) =** after entering the last data point.

Step 2   Once the data is entered, use the calculator to find the mean (average) of the data.

• Return to the **HOME** screen by pressing 2nd and MODE .

• Go to the **LIST** menu.

TI-83+   Press 2nd and STAT . Press ▶ twice and select mean(from the **MATH** submenu.

TI-73   Press 2nd and LIST . Press ▶ twice and select mean(from the **MATH** submenu.

• The **MEAN** function divides the sum of all the data by the number of data points in **L1**.

TI-83+   Press [2nd] [1] and [ )]. Press [ENTER].

TI-73   Press [2nd] [LIST], select [L1], and press [ )]. Press [ENTER].

```
NAMES OPS MATH
1:min(
2:max(
3▓mean(
4:median(
5:sum(
6:prod(
7↓stdDev(
```

The mean of the data set is _____.

Step 3   In order to find the median (middle number) of the data, use the same menu options as you did for the mean.

• Return to the **LIST** menu.

TI-83+   Press [2nd] and [STAT].

TI-73   Press [2nd] and [LIST].

• Press [▶] twice and select median(from the **MATH** submenu.

• The **MEDIAN** function finds the average of the two middle numbers in **L1**.

TI-83+   Press [2nd] [1] and [ )]. Press [ENTER].

TI-73   Press [2nd] [LIST], select [L1] and press [ )]. Press [ENTER].

```
NAMES OPS MATH
1:min(
2:max(
3:mean(
4▓median(
5:sum(
6:prod(
7↓stdDev(
```

The median of the data set is _____.

Step 4   To find the mode of the data, follow the directions for the calculator you are using.

TI-83+   Sort the data from least to greatest in order to find the mode (most frequent value).

• Return to the **LIST** menu. Press [2nd] and [STAT].

• Press [▶] once and select SortA(from the **OPS** submenu.

```
NAMES OPS MATH
1▓SortA(
2:SortD(
3:dim(
4:Fill(
5:seq(
6:cumSum(
7↓▲List(
```

     Saxon Math Course 3

• Now press ⏺2nd ⏺1 and ⏺). This function will sort the data points in **L1** in ascending order. Press ⏺ENTER. The screen should display **Done**.

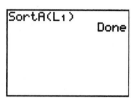

• To see the sorted data, return to the **Edit** menu by pressing ⏺STAT then selecting ⌷EDIT...⌷. If you press ⊖ to scroll, you can see how the data points increase in value.

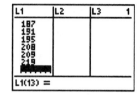

• Now find the mode of the data in **L1** by pressing ⊕ or ⊖ and look for the one data point that repeats the most.

TI-73 Use the same menu options as we did for the mean and median to find the mode (most frequent value).

• Return to the **LIST** menu by pressing ⏺2nd and ⏺LIST.

• Press ⏺▶ twice and select ⌷mode(⌷ from the **MATH** submenu.

• Now press ⏺2nd ⏺LIST, select ⌷L1⌷, and press ⏺). This function finds the value that repeats the most in **L1**. Press ⏺ENTER.

The mode of the data set is _____.

Practice

Use your graphing calculator to answer the following questions.

1. Roll two number cubes 25 times and record the sums of the faces showing up. Enter your data into **L1** in the calculator. Then use the functions in the **MATH** submenu to find the mean, median, and mode of your data.

    a. mean   _____

    b. median   _____

    c. mode   _____

d. Which measure of central tendency best represents the typical sum from a roll of the number cubes? Explain your answer.

2. A car dealership's monthly sales are displayed on the table. Use the functions in the **MATH** submenu to find the mean, median, and mode of the data.

a. mean _____

b. median _____

c. mode _____

d. Which measure of central tendency would distort the fact that the dealership did not sell more than 1000 vehicles for most months? Explain your answer.

Month	Number of Vehicles Sold
Jan	550
Feb	450
Mar	520
Apr	620
May	620
June	750
July	805
Aug	885
Sep	3865
Oct	360
Nov	250
Dec	2345

Saxon Math Course 3

• The Coordinate Plane

New Calculator Functions: PLOT , ZOOM , and TRACE

Example: Graph the pairs of numbers from this table on a coordinate plane.

x	y
–2	–1
–1	0
0	1
1	2

Demonstration

Step 1 To graph the pairs of numbers from the table in the calculator, first enter them into **L1** and **L2**.

• Go to the **STAT** menu.

TI-83+ Press STAT and select EDIT....

TI-73 Press LIST .

• Enter the *x*-values from example 2 into **L1** by pressing (−) 2 and ENTER , pressing (−) 1 and ENTER , and so on. Your screen should display **L1(5) =** after entering the last *x*-coordinate.

• Enter the *y*-values from example 2 into **L2** by pressing ▶, (−) 1 and ENTER , pressing 0 and ENTER , and so on. Your screen should display **L2(5) =** after entering the last *y*-coordinate.

Step 2 Once the pairs of numbers are entered, set the calculator to plot them on the coordinate plane.

• Press 2nd and Y= . Select Plot1.

• Press ENTER to turn **Plot1** On. Select |·⋅⋅ under TYPE: by pressing ▼ once then ENTER .

Name _____

- To ensure that the pairs of data will be plotted in the correct order, be sure that **L1** is entered in ⬚Xlist:⬚ and **L2** is entered in ⬚Ylist:⬚.

  If not, set ⬚Xlist:⬚ to **L1** and ⬚Ylist:⬚ to **L2**:

TI-83+   Press ⬇ once for ⬚Xlist:⬚, then 2nd and 1 .
Press ⬇ once more for ⬚Ylist:⬚, then 2nd and 2 .

TI-73   Press ⬇ once for ⬚Xlist:⬚, then 2nd LIST and 1 . Press ⬇ once more for ⬚Ylist:⬚, then 2nd LIST and 2 .

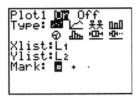

Step 3   Now we can see what the pairs of numbers look like on the coordinate plane.

- Press ZOOM and select ⬚ZStandard⬚.

- To make sure the pairs of numbers are correctly graphed, press TRACE and ▶ or ◀. The screen should display **X=** and **Y=** at the bottom for each pair.

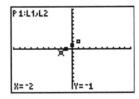

     Saxon Math Course 3

Graph the pairs of numbers on a coordinate plane.

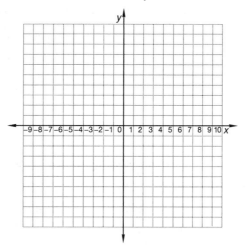

Practice

Use your graphing calculator to answer the following questions.

1. Pair up with another student with some graph paper and two calculators. Without showing the other student, plot the vertices of a polygon on graph paper. Take turns calling out one vertex (pair of numbers) at a time from each polygon. At each turn enter the other student's x-value into **L1**, enter the y-value into **L2**, and then take a guess of what the polygon is by graphing the pairs on the calculator. Call out the number pairs one at a time until one of you guesses the name of the other's polygon.

    a. If this were a game of simply who could guess more quickly, what would be a disadvantage?

    _____

    _____

    b. What could you do to give yourself more of an advantage?

    _____

    _____

c. If time permits, each student may draw another polygon on graph paper with the advantages and disadvantages of the game in mind. Take turns calling out pairs of numbers and guessing what polygon is being drawn. Did analyzing a disadvantage and advantage of the activity affect the next round of play? Explain.

2. The path of an airplane is shown in the given table as ordered pairs, where the *x*-value represents the horizontal distance from your location at (0, 0) and the *y*-value represents the vertical distance from the ground. Which line graph best represents the path of the airplane?

x	y
−1	1
0	3
1	5
2	7

A
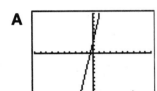

C
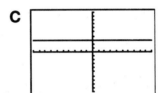

B
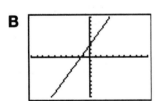

D
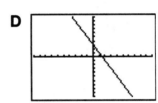

• Adding and Subtracting Mixed Numbers

New Calculator Functions: ▸Frac or 🔲 ⅗ 🔲

Example: The carpenter cut $15\frac{1}{2}$ inches from a 2-by-4 that was $92\frac{5}{8}$ inches long. How long was the resulting 2-by-4?

Demonstration

To calculate the sum or difference of mixed numbers, follow the directions for the calculator you are using.

Step 1 Enter the mixed numbers into the calculator as an addition of the whole number with the fraction.

TI-83+ Enter the subtraction from example 5 by pressing 🔲 (92 🔲 + 🔲 5 🔲 ÷ 🔲 8 🔲) 🔲 and 🔲 − 🔲 then 🔲 (🔲 15 🔲 + 🔲 1 🔲 ÷ 🔲 2 🔲) 🔲. Press ENTER.

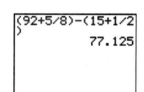

TI-73 Enter the subtraction from the example 5 by pressing 🔲 (🔲 92 🔲 + 🔲 ⅗ 5 🔲 ⊖ 8 🔲 ▸ 🔲) 🔲 and 🔲 − 🔲 then 🔲 (🔲 15 🔲 + 🔲 ⅗ 🔲 1 ⊖ 2 ▸ 🔲) 🔲. Press ENTER.

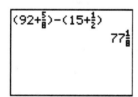

Step 2 Notice the answer is a decimal number. To convert the solution to a mixed number, separate the whole number from the decimal part.

TI-83+ • Press MATH and then ▸ once to go to the **NUM** submenu. Select fPart(. This function will separate the fraction part from the decimal number. Press 2nd ANS) then ENTER. The screen should display only the decimal part of the answer.

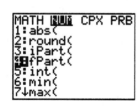

Step 3 Now that you have separated the different parts from the solution, convert the decimal part to a fraction.

TI-83+ • Press MATH and select ▸Frac. Press ENTER. The mixed number answer to example 5 is the whole number of our solution with the fraction of the decimal number.

How long was the resulting 2-by-4? _____

Practice

Use your graphing calculator to answer the following questions.

1. The table shows the ingredients for a low-fat pasta salad with chicken. Collaborate with 2 other students. Find the total amount of each of the following ingredients that you would need to prepare the pasta salad for all 3 students in your group. For each step, discuss what mathematical operation you will need to use (addition or subtraction) and then enter an expression into the graphing calculator.

Low-Fat Pasta Salad with Chicken	
(Makes 1 serving each)	
$1\frac{1}{3}$ c.	spiral wheat pasta
$2\frac{2}{3}$ Tbsp	olive oil
$\frac{1}{4}$ Tbsp	Italian seasoning
$\frac{3}{16}$ lb.	low-fat, skinless chicken breast
$1\frac{1}{2}$ c.	steamed vegetables
$\frac{3}{4}$ oz.	reduced-fat feta cheese

a. How much pasta, chicken, and feta will you need?

b. How much olive oil and Italian seasoning will be left in the pasta salad if only one of your group members has one serving?

c. Suppose you already had $2\frac{1}{3}$ cups of vegetables. How much more would you need in order to make the pasta salad?

2. The box used to ship your calculator weighed $5\frac{1}{16}$ pounds. If the shipping materials weighed $1\frac{1}{5}$ pounds, how much does your calculator weigh?

3. A cereal box contains $14\frac{2}{3}$ ounces of grains, $\frac{1}{8}$ ounce of plastic wrap, and $2\frac{3}{4}$ ounces of cardboard. What is the total weight of the cereal box?

   A $11\frac{19}{24}$ oz   **B** $13\frac{7}{24}$ oz   **C** $17\frac{5}{12}$ oz   **D** $17\frac{13}{24}$ oz

Saxon Math Course 3

Name _____

• Powers and Roots

New Calculator Functions: x^2 and $\sqrt{}$

Example: Simplify

    a. $\sqrt{144}$ **b.** $\sqrt[3]{27}$

Demonstration

Step 1 To find the solution to example 5a, use the **square root** function of the calculator.

    • Press [2nd] and [x^2]. Enter 144 and press [ )]. Press [ENTER]. The screen should display **12**.

    • To perform the inverse operation of the square root, square your answer by pressing [x^2] and [ENTER]. The screen should display **144**.

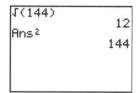

    The $\sqrt{144}$ is _____.

Step 2 To find the solution to example 5b, use the **cubed root** function of the calculator.

    • Press [MATH] and select [$\sqrt[3]{\ }($]. Enter 27 and press [ )]. Press [ENTER]. The screen should display **3**.

    • To perform the inverse operation of the cubed root, cube your answer by pressing [MATH] and selecting [3]. Press [ENTER]. The screen should display **27**.

    The $\sqrt[3]{27}$ is _____.

Step 3 To find the solution to $\sqrt[5]{32}$, use the **xth root** function of the calculator.

    • First press [5] for the 5th root. Then return to the **MATH** submenu by pressing [MATH]. Select [$\sqrt[x]{\ }$], enter 32, and press [ENTER]. The screen should display **2**.

- To perform the inverse operation of the 5th root, raise your answer to the 5th power by pressing [^] and [5]. Press [ENTER]. The screen should display **32**.

The $\sqrt[5]{32}$ is _____.

Practice

Use your graphing calculator to answer the following questions.

1. A paper mill makes poster board in square sheets. Each sheet has a total area of 1225 square inches. The mill has decided to change its design and manufacture rectangular poster board, charging the same price. However, the machinery requires that the amount decreased on one dimension be added to the other dimension. The total area of the rectangle will be 1125 square inches. With a partner and a graphing calculator, find the dimensions of the rectangular sheets.

   a. What are the dimensions for one of the current square sheets?

   b. What are the dimensions for one of the new rectangular sheets?

   c. Are the new rectangular poster boards a good deal for the consumer? Explain.

2. Damaris is calculating the distance from 1st to 3rd base in a baseball diamond. Unfortunately she does not have a calculator and cannot simplify her answer of $\sqrt{16200}$ ft. She knows that the distance should be approximately 127 ft. Describe how Damaris could verify her answer without a calculator. Is her approximation correct?

Saxon Math Course 3

3. Julian picks a positive number, raises it to a positive power, and gets an answer of 81. If the number he picks is less than 9, which of the following represents that number?

A $\sqrt{81}$

B $\sqrt[3]{81}$

C $\sqrt[4]{81}$

D Not Possible

• Multiplying and Dividing Mixed Numbers

New Calculator Functions: Frac or $\left(A\frac{b}{c}\leftrightarrow\frac{d}{e}\right)$

Example: Simplify

a. $2\frac{1}{3} \times 1\frac{1}{2} =$ **b.** $1\frac{2}{3} \div 2\frac{1}{2} =$

Demonstration

Step 1 To calculate the product of mixed numbers, first convert them to improper fractions in the calculator.

• Enter the first mixed number from example 1a.

TI-83+ Press 2 **+** 1 **÷** 3 then **ENTER**. Press **MATH** and select Frac then press **ENTER**. The screen should display **7/3**.

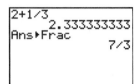

TI-73 Press 2 **+** 1 **÷** 3 then **ENTER**. Press $\left(A\frac{b}{c}\leftrightarrow\frac{d}{e}\right)$ and **ENTER**. The screen should display $\frac{7}{3}$.

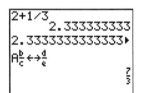

• Enter the second mixed number from example 1a.

TI-83+ Press 1 **+** 1 **÷** 2 then **ENTER**. Press **MATH** and select Frac, then press **ENTER**. The screen should display **3/2**.

TI-73 Press 1 **+** 1 **÷** 2 then **ENTER**. Press $\left(A\frac{b}{c}\leftrightarrow\frac{d}{e}\right)$ and **ENTER**. The screen should display $\frac{3}{2}$.

Step 2 Now enter both improper fractions to find the product of the mixed numbers.

• Press **(** 7 **÷** 3 **)** and **×** then **(** 3 **÷** 2 **)**. Press **ENTER**. The screen should display **3.5**.

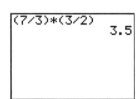

 Saxon Math Course 3

Step 3   Convert the answer to a mixed number.

• Use the calculator to convert the decimal number.

TI-83+   Enter 0.5 (the decimal part of our answer), press
MATH, select ⌊frac⌋, then press **ENTER**. The screen
should display **1/2**. Our mixed number answer is
the whole number **3** and **1/2**.

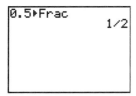

```
0.5▶Frac
              1/2
```

TI-73   Enter 3.5. Press ⌈F↔D⌋ and **ENTER**. The screen should
display $3\frac{1}{2}$.

```
3.5▶F↔D        3½
```

$2\frac{1}{3} \times 1\frac{1}{2} = $ _____

Follow steps 1–3 to find the quotient in example 1b.

$1\frac{2}{3} \div 2\frac{1}{2} = $ _____

Practice

Use your graphing calculator to answer the following questions.

1. With a partner, measure (in inches) the dimensions of a square or rectangular
   object in the classroom. Then find the area of the object using the graphing
   calculator. Express your measurements in mixed numbers.

   a. Write an equation for the product of the dimensions (in improper fraction form).

   b. Suppose you were to divide your object into $4\frac{1}{8}$ parts lengthwise. What would be
   the length of each part (in mixed number form)?

   c. Will the segments be easy to measure? Explain.

2. A mother, father, and child share a meal. The child eats $\frac{2}{3}$ as much food as one adult. If the recipe for one adult serving calls for two and a fourth cups of milk, how much milk is used for the meal? Explain you answer.

3. A cup filled with liquid weighs $1\frac{4}{5}$ pounds. If the cup weighs $4\frac{1}{2}$ times the weight of the liquid, how much does the liquid weigh to the nearest fraction of a pound?

Name _____

• Scientific Notation of Large Numbers

New Calculator Functions: Sci and EE

Example: Earth's average distance from the Sun is 93,000,000 miles. Write 93,000,000 in scientific notation.

Demonstration

Step 1 First, set the calculator so that all numbers are in scientific notation.

- Press MODE and select Sci by pressing ▶ once and ENTER.

- Return to the **HOME** screen by pressing 2nd and MODE. Enter the number from example 1, 93000000, without commas, and then press ENTER.

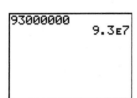

- Convert our calculator answer to scientific notation by recognizing that the **E** symbolizes ×**10** and that **7** is the exponent of 10.

Write 93,000,000 in scientific notation. _____

Example: When Brigita finished the 1500 meter run she announced that she had run "one point five times ten to the sixth millimeters." Write that number **a** in scientific notation and **b** in standard form.

Step 2 To convert a number in scientific notation back to standard form, we must reset the calculator to **Normal Mode**.

- Press MODE and ENTER. **Normal** should be highlighted in black.

- Return to the **HOME** screen by pressing 2nd and MODE.

• Convert the scientific number 1.5×10^6 in example 2 to standard form.

```
1.5E6
        1500000
```

TI-83+   Enter 1.5, press [2nd] [,], then enter 6. Press [ENTER].

TI-73   Enter 1.5, press [2nd] [^], then enter 6. Press [ENTER].

Write Brigita's distance **a.** in scientific notation and **b.** in standard form.

a. _____

b. _____

___

Practice

Use your graphing calculator to answer the following questions.

1. Arrange your desks in groups of 2 or 3, so that they are side-by-side. Then cut a strip of adding machine tape equal to the full length of your desktops. Draw a number line along the middle and mark 0 on the left end of the tape. When your teacher calls out a number, convert it to standard form or scientific notation and position both on your number line. As new numbers are announced, you may need to erase and reposition the other numbers so that they are all accurately placed on your number line.

   a. What factor made placing the numbers on the number line challenging?

   b. Which numbers required you to erase and reposition the other numbers? Explain.

c. Did you have to reposition the other numbers when the last two numbers were announced? Explain.

2. In 2005, the world's wealthiest person was worth $51,500,000,000. What is the correct scientific notation for that person's wealth?

3. At the end of a particular day, Stock XYZ's share value was $31.29, down $0.31, at a trading volume of 1.18×10^6. Which of the following is the correct standard form for the stock's trading volume?

    A 75.6

    B 1,180,000

    C 11,800,000

    D Not Here

4. Josh writes down two numbers with an inequality symbol.

$$5.67 \times 10^7 < 8.81 \times 10^6$$

Is he correct? Explain.

• Subtracting Integers

New Calculator Function: ((−))

Example: Change each subtraction to addition and find the sum.

      a. (−3) − (+2)   **b.** (−3) − (−2)

Demonstration

Step 1   When subtracting a positive number, use the **addition** and **subtraction** functions.

        • Enter the subtraction (−3) − (+2) from example 1a. Press (( ) ((−)) (3) ( )) and (−) then (( ) (2) ( )). Press (ENTER).

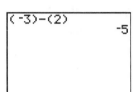

        • Enter the added opposite (−3) + (−2) by using the **negative** function. Press (( ) ((−)) (3) ( )) and (+) then (( ) ((−)) (2) ( )). Press (ENTER).

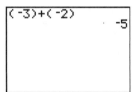

    (−3) − (+2) = _____   (−3) + (−2) = _____

Step 2   When subtracting a negative number, use the **subtraction** and **negative** functions.

        • Enter the subtraction (−3) − (−2) from example 1b. Press (( ) ((−)) (3) ( )) and (−) then (( ) ((−)) (2) ( )). Press (ENTER).

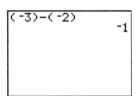

        • Enter the added opposite (−3) + (+2) by pressing (( ) ((−)) (3) ( )) and (+) then (( ) (2) ( )). Press (ENTER).

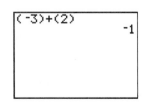

    (−3) − (−2) = _____   (−3) + (+2) = _____

    Saxon Math Course 3

Name _____

Practice

Use your graphing calculator to answer the following questions.

1. Model the addition and subtraction of integers on the number line. Draw a
    number line with a scale of 2 from –20 to 20. Start at 0 and use arrows to
    denote each addition or subtraction. (See the sample number line.) Use the
    graphing calculator to find the answer as your teacher calls out each integer.

    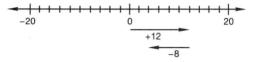

    a. With which integer does all the addition and subtraction end? _____

    b. Explain why that integer is the final answer.

    c. If time permits, create your own list of addition and subtraction that ends with the
    same integer you started. Have another student draw your list on a number line
    as you call out the integers. Could you use an odd number of integers and get the
    same result? Explain.

2. The new AquaPlane can fly like an airplane to an altitude of 35,000 ft in the air
    and sink like a submarine to a depth of –3500 ft.

    a. Write a subtraction expression for the AquaPlane's total range in altitude.

    b. Write the added opposite of your expression.

    c. What is the AquaPlane's range in altitude?

       A 3500 ft

       B 31,500 ft

       C 35,000 ft

       D 38,500 ft

3. Cobey writes the following on the board.

$$9 - (-9) = (9) + (+9)$$

Is she correct? Explain.

Saxon Math Course 3

Name _____

• Multiplying and Dividing Integers

Calculator Function: (−)

Example: Simplify

    a. (−10)(−5)   **b.** (+10)(−5)

    c. $\frac{-10}{-5}$ **d.** $\frac{10}{-5}$

Demonstration

Step 1   Multiply integers in the calculator using the **negative** function.

    • Enter the multiplication from example 1a. Press

    ((−) 1 0) then ((−)

    5) . Press ENTER .

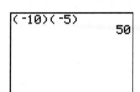

    • Enter the multiplication from example 1b. Press (

    1 0) then ((−) 5) .

    Press ENTER .

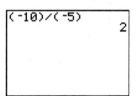

    (−10)(−5) _____   (+10)(−5) _____

Step 2   To divide integers, follow the directions for the calculator you are using.

TI-83+   Divide using the **division** function.

    • Enter the division $\frac{(-10)}{(-5)}$ from example 1c.

    Press ((−) 1 0) and ÷
    then ((−) 5) . Press ENTER .

    • Enter the division $\frac{(10)}{(-5)}$ from example 1d.

    Press (1 0) and ÷ then
    ((−) 5) . Press ENTER .

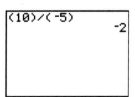

TI-73   Divide using the **stacked fraction** function.

- Enter the division $\frac{-10}{-5}$ from example 1c.

  Press ⬜%⬜ ⬜(−)⬜ ⬜1⬜ ⬜0⬜ and ⬇ once
  then ⬜(−)⬜ ⬜5⬜. Press ⬜ENTER⬜.

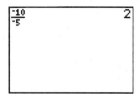

- Enter the division $\frac{10}{-5}$ from example 1d.

  Press ⬜%⬜ ⬜1⬜ ⬜0⬜ and ⬇ once then
  ⬜(−)⬜ ⬜5⬜. Press ⬜ENTER⬜.

$\frac{(-10)}{(-5)}$ _____ $\frac{(10)}{(-5)}$ _____

Practice

Use your graphing calculator to answer the following questions.

1. The object of this game is to become the first integer millionaire. Arrange yourselves in groups of 2 or more students. Shuffle the cards and deal them face down to the players. In this game, every black number card represents a positive number, and every red number card represents a negative number. Also, a number card indicates multiplication, and a face card indicates division by 10 (positive or negative).

Type of Card	Do	By
Black 2–10	×	pos. #
Red 2–10	×	neg. #
Black J, Q, K	÷	+10
Red J, Q, K	÷	−10
Black Ace	Keep	
Red Ace	Change	

An ace counts as one, so a black ace keeps the sign, and a red ace switches the sign.

The player to the left of the dealer turns up his/her top card and places it in the middle to start the game. Each player should enter this number into his/her calculator. The next player to the left turns up one card and places it on top of the discard pile. Everyone then multiplies or divides the value of his card with the value recorded in the calculator. If a division creates a decimal number, everyone will enter only the integer part into the calculator or 1, whichever is greater.

The game ends with the first player who turns up a card that multiplies to a product above +1,000,000. If no one gets to one million and all have run out of cards, then the discard pile is reshuffled and a new round begins.

       Saxon Math Course 3

a. What were the challenges of multiplying to a million?

b. Was there a connection between the signs of the cards and the resulting product or quotient? Explain.

2. David and two partners own a lemonade stand. In addition to helping him run the business, the partners equally share the expenses of $99. He shows you the following expression for calculating each partner's share.

$$\frac{-99}{+3}$$

Why does it make sense for David to use his expression instead of $\frac{+99}{-3}$ in the real-world?

A −99 represents income

B 3 represents people

C positive divided by a negative is positive

D negative divided by a positive is negative

3. Without doing the calculation, identify the answer to (−3)(+4)(+5)(−126)(−1) as positive or negative. Explain your answer.

• Graphing Functions

New Calculator Functions: Y= and TABLE

Example: An online auction service charges a 30¢ listing fee plus 6% of the selling price for items up to $25. The total service charge (*c*) is a function of the selling price (*p*) of an item.

$$c = 0.30 + 0.06p$$

a. Use a calculator to generate a table of pairs of values for the selling price and service charge.

b. Graph the function on axes like the ones below.

c. Is the relationship a proportion? How do you know?

Demonstration

Step 1 To generate a table for example 3a from the function $c = 0.30 + 0.06p$ first enter it into the calculator.

• Go into **Y= Editor**.

TI-83+ Press Y= . Press 0 · 3 0 and + then 0 · 0 6 . Now press X,T,θ,n to assign the input variable to the function.

```
Plot1 Plot2 Plot3
\Y1■0.30+0.06X
\Y2=
\Y3=
\Y4=
\Y5=
\Y6=
\Y7=
```

TI-73 Press Y= . Press 0 · 3 0 and + then 0 · 0 6 . Now press x to assign the input variable to the function.

```
Plot1 Plot2 Plot3
\Y1■0.30+0.06X
\Y2=
\Y3=
\Y4=
```

  Saxon Math Course 3

Name _____

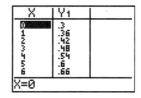

- To see the table of values for the function, press 2nd and GRAPH.

Generate a table from the function $c = 0.30 + 0.06p$ in the space provided.

Step 2 For example 3b, graph $c = 0.30 + 0.06p$ by using the **ZOOM** function of the calculator.

- Press ZOOM and select ZStandard. The screen should display a line that is hard to make out.

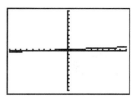

- In this case, zoom to the part of the graph that is easier to see. Press ZOOM and select ZoomFit. The screen should now display a line that is much clearer.

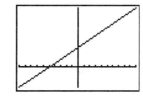

Graph the function $c = 0.30 + 0.06p$ on axes like the ones shown in Example 3b.

Step 3 Find out if the relationship is proportional for example 3c through one of two methods.

- See if the graph goes through the origin by pressing TRACE. The screen should display **X=0** and **Y=.3** at the bottom.

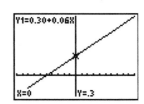

- Or look at the pair of values near the origin by pressing 2nd and TABLE. Press ⌃ or ⌄ so that the screen displays **X=0** and **Y1=.3**.

Is there a proportional relationship between the variables in the function $c = 0.30 + 0.06p$? Explain your answer.

Practice

Use your graphing calculator to answer the following questions.

1. Pair up with another student to generate a function, table, and graph for the following scenario. You and your friend decide to earn money by mowing lawns over the summer. You know that a local company charges a flat fee of $15 plus $10 per hour of service. Determine the flat fee and hourly rate you will charge customers in order to compete with the local company. Keep in mind that you will have to pay an unspecified amount for gas, mower rental, and wages for you and your friend. Then write a function for the income of your business, generate a table and graph using the graphing calculator.

 Saxon Math Course 3

a. What variable letters did you use in your function? Explain what they represent.

b. Will the function you generated for the mowing business compete with the local company? Explain.

c. In what ways will your function not be competitive?

2. Consider the function $y = -3x + 5$.

a. Generate a table from the function.

b. Generate a graph from the function.

c. Is there a proportional relationship between the variables? Explain your answer.

3. The function $p = 0.5c$ represents the number of cups in a pint, where c is cups and p is pints.

a. Which of the following tables contains the pairs of values for this function?

A

c	1	2	3	4	5
p	5	10	15	20	25

B

c	1	2	3	4	5
p	0.05	0.10	0.15	0.20	0.25

C

c	1	2	3	4	5
p	0.5	0.5	0.5	0.5	0.5

D

c	1	2	3	4	5
p	0.5	1.0	1.5	2.0	2.5

b. Is there a proportional relationship between pints and cups? Explain.

c. Does the function accurately portray a real-world relationship? Explain.

Saxon Math Course 3

• Transformations

New Calculator Function: 〰️

Example: Draw triangle *ABC* as illustrated on a coordinate plane. Then draw its image △*A'B'C'* translated 7 units to the right and 8 units down, or (7, –8). Label the coordinates of △*A'B'C'*.

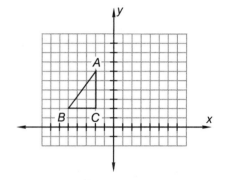

Demonstration

Step 1 In order to translate △*ABC*, first enter its coordinates into the **STAT** function so that the calculator can connect the vertices of the figure.

- Go to the **STAT** menu.

TI-83+ Press [STAT] and select [Edit...].

TI-73 Press [LIST].

- Enter the *x*-coordinates from problem 6 into **L1** by pressing [(–)] [2] and [ENTER], and so on. Repeat the *x*-coordinate for point *A* at the end of the list. Your screen should display **L1(5) =**.

- Enter the *y*-coordinates into **L2** by pressing [▶] [6] and [ENTER], and so on. Be sure to repeat the *y*-coordinate for point *A* at the end of the list. Your screen should display **L2(5) =**.

Step 2 Now calculate the coordinates of the translated image △*A'B'C'* and enter them into the **STAT** function.

- Press [2nd] and [MODE] to return to the **HOME** screen.

- Problem 6 says to translate the image 7 units to the right. Use the calculator to store the *x*-coordinates of the translated image into **L3**.

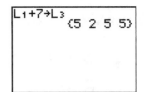

TI-83+   Press [2nd] [1]. Then press [+] [7] and [STO•]. Press [2nd] [3] and [ENTER].

TI-73   Press [2nd] [LIST] [1]. Then press [+] [7] and [STO•]. Press [2nd] [LIST] [3] and [ENTER].

- Problem 6 also says to translate the image 8 units down. Use a similar function to store the *y*-coordinates of the translated image into **L4**.

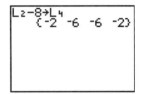

TI-83+   Press [2nd] [2]. Then press [−] [8] and [STO•]. Press [2nd] [4] and [ENTER].

TI-73   Press [2nd] [LIST] [2]. Then press [−] [8] and [STO•]. Press [2nd] [LIST] [4] and [ENTER].

Write down the calculations of the coordinates that you entered into the calculator.

Step 3   Once the coordinates of the figure and its translated image are entered, set the calculator to plot them on the coordinate plane.

- Press [2nd] and [Y=]. Select [Plot1]. Press [ENTER] to turn **Plot1** [On]. Select [⁀] under [Type:] by pressing [▼] once, [▶] once, then [ENTER]. To ensure that the coordinates from the original figure will be plotted, be sure that **L1** is entered in [Xlist:] and **L2** is entered in [Ylist:].

TI-83+   If they are not, press [▼] once for [Xlist:] then [2nd] [1]. Press [▼] once more for [Ylist:] then press [2nd] [2].

    Saxon Math Course 3

TI-73   If they are not, press ⌄ once for Xlist:,
2nd LIST 1 . Press ⌄ once more for
Ylist: then 2nd LIST 2 .

- Press 2nd and Y= . Select Plot2.... Press
  ENTER to turn **Plot2** On. Select 〰 under Type: by
  pressing ⌄ once, ▶ once, then ENTER . To ensure
  that the coordinates from the translated image will be
  plotted, be sure that **L3** is entered in Xlist: and **L4** is
  entered in Ylist:.

TI-83+   If they are not, press ⌄ once for Xlist: and 2nd
  3 . Press ⌄ once more for Ylist: and 2nd
  4 .

TI-73   If they are not, press ⌄ once for Xlist: and
  2nd LIST 3 . Press ⌄ once more for
  Ylist: and 2nd LIST 4 .

Step 4   Now graph the original image and its translated image on a coordinate
plane.

- Press ZOOM and select ZStandard.

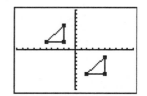

- To make sure the pairs of numbers for the
  original figure are graphed correctly, press
  TRACE and ▶ or ◀. The screen should display
  X= and **Y=** at the bottom for each pair.

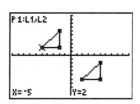

Name _____

- To make sure the pairs of numbers for the translated image are graphed correctly, press ⊙ then ▶ or ◀. The screen should display **X=** and **Y=** at the bottom for each pair.

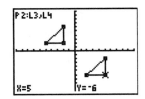

Graph the original image and its translated image on the coordinate plane provided below.

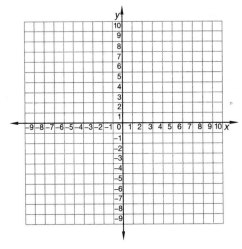

Write down the coordinates of the translated image.

_____

Practice

Use your graphing calculator to answer the following questions.

1. Rectangle *WXYZ* is translated 5 units to the right and 6 units up.

   a. Write down the coordinate calculations you enter into the calculator.

   _____

   _____

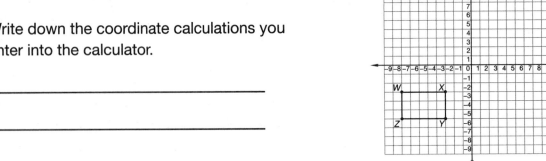

Saxon Math Course 3

b. Graph the translated image of rectangle *WXYZ* on the coordinate plane provided below.

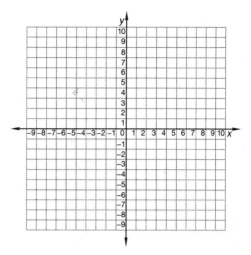

c. Use the calculator to find the coordinate of point *Z'*. _____

2. Justin is a graphics programmer for an animated cartoon. His task is to move the given figure to Quadrant II of the coordinate plane. However, the image must be the same distance from the origin as the original figure.

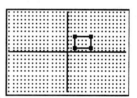

a. Write down the coordinate calculations you enter into the calculator.

b. Graph the translated image on the coordinate plane provided below.

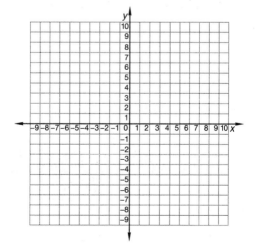

c. What other type of transformation did Justin perform?

• Scientific Notation for Small Numbers

Calculator Functions: Sci, ^ , and (−)

Example: Write this number in standard form.

$$1.5 \times 10^{-3}$$

Demonstration

Step 1   To convert a number in scientific notation to standard form, use the **power** function.

- Press **MODE** and **ENTER**. **Normal** should be highlighted in black. If it is not, press **ENTER**.

- Return to the **HOME** screen by pressing **2nd** and **MODE**.

- Enter the number from example 6, **1.5×10^{-3}**, by pressing **1** **.** **5** **×** **1** **0** and **^** **(−)** **3**. Press **ENTER**.

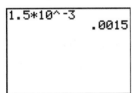

Write 1.5×10^{-3} in standard form.

Example: The diameter of a red blood cell is about 0.000007 meters. Write that number in scientific notation.

Step 2   To convert a number in standard form to scientific notation, we must set the calculator to **Scientific Mode**.

- Press **MODE** and select Sci by pressing ▶ once and **ENTER**.

TI-83+

TI-73

Saxon Math Course 3

- Return to the **HOME** screen by pressing [2nd] and [MODE]. Enter the number from example 7, 0.000007, then press [ENTER].

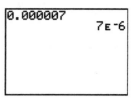

- We can write our calculator answer in scientific notation by recognizing that **E** symbolizes **×10** and that **−6** is the exponent of 10.

Write 0.000007 in scientific notation. _____

_____

Practice

Use your graphing calculator to answer the following questions.

1. Arrange your desks in groups of 2 or 3 so that they are side-by-side. Then cut a strip of adding machine tape equal to the full length of your desktops. Draw a number line along the middle, mark 0 on the left, and mark 1 on the right end of the tape. When your teacher calls out a number, convert it to standard form or scientific notation and position both on your number line. As new numbers are announced, you may need to erase and reposition the other numbers so that they are all accurately placed on your number line.

   a. What made placing the numbers on the number line challenging?

   b. Which numbers forced you to erase and reposition the other numbers? Explain.

   c. Recall the similar problem in Graphing Calculator Activity 6. What is the main difference between that problem and today's problem?

2. A nanosecond was first used to express the speed of a computer's processing. Today's computers can process bytes of information much faster than a nanosecond, but in the beginning the bit rate was 0.000000001. Which of the following is the correct scientific notation for a nanosecond?

   A 1.0×10^{-9}

   B 1.0×10^{-10}

   C 1.0×-9

   D Not Here

3. A particulate is a small airborne particle like dust. Fine particulates are less than 2.5 microns, where each micron is equivalent to 1.0×10^{-6} meter. Express 2.5 microns as meters in scientific notation and in standard form.

4. Gavin writes down two numbers with an inequality symbol.

$$8.81 \times 10^{-7} < 5.67 \times 10^{-6}$$

Is he correct? Explain.

• Collect, Display, and Interpret Data

New TI-73 Calculator Functions: ⌊ₙ⌋, ⊘ , and ⊞

Example: One hundred eighth graders were surveyed and asked what extra-curricular activities they planned to be involved with in high school. The students were given several choices and could select as many as they wanted. The results are shown below.

music	卌 卌 卌 卌 卌 II
sports	卌 卌 卌 卌 I
service clubs	卌 卌 卌 卌 卌 卌 III
student government	卌 卌 卌
other	卌 卌 III

Choose a type of graph (bar graph, histogram, or circle graph) for representing these data, and explain your choice.

Demonstration:

This demonstration is for the **TI-73** graphing calculator. Once you have decided on the type of graph to use, follow the appropriate steps.

Step 1 To graph the data from the frequency chart, first enter it into the **LIST** function.

• Press (LIST). Label the categories from problem 7 with 1 through 5 and enter them into **L1** by pressing (1) and (ENTER), (2) and (ENTER), and so on. Your screen should display **L1(6) =** after entering the last label.

• Enter the quantities for each category into **L2** by pressing (▶)(2)(7) and (ENTER), (2)(1) and (ENTER), and so on. The screen should display **L2(6) =** after entering the last quantity.

Step 2 Set the calculator to graph the labels and quantities as a **bar graph** by using the **PLOT** menu.

- Press [2nd] and [Y=]. Select ⌜Plot1...⌟. Press [ENTER] to turn **Plot1** ⌜On⌟.

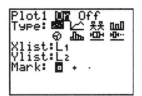

- Select ⌜lₙl⌟ under ⌜Type:⌟ by pressing ⌄ once, (▸) three times, then [ENTER].

- To ensure that the pairs of data will be plotted in the correct order, be sure that **L1** is entered in ⌜CategList:⌟ and **L2** is entered in ⌜DataList1:⌟.

    If they are not, press ⌄ once for ⌜CategList:⌟, [2nd] [LIST], and [1]. Press ⌄ once more for ⌜DataList1:⌟, [2nd] [LIST] and [2].

- Be sure that ⌜Vert⌟ and ⌜1⌟ are highlighted in black at the bottom of the screen. If they are not, press ⌄ three times from ⌜DataList1:⌟ and press [ENTER]. Then press (▸) twice and then [ENTER].

    Or set the calculator to graph the labels and quantities as a **circle graph** by using the **PLOT** menu.

- Press [2nd] and [Y=]. Select ⌜Plot1...⌟. Press [ENTER] to turn **Plot1** ⌜On⌟.

- Select ⊘ under ⌜Type:⌟ by pressing ⌄ once, (▸) four times, then [ENTER].

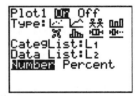

- To ensure that the pairs of data will be plotted in the correct order, be sure that **L1** is entered in ⌜CategList:⌟ and **L2** is entered in ⌜DataList:⌟.

Saxon Math Course 3

If they are not, press ⊙ once for `CategList:`, `2nd` `LIST`, and `1`. Press ⊙ once more for `DataList:`, `2nd` `LIST`, and `2`.

- Be sure that `Percent` is highlighted in black at the bottom of the screen. If not, press ⊙ once from `DataList:`, `▶` once, then `ENTER`.

Or set the calculator to graph the labels and quantities as a **histogram** by using the **PLOT** menu.

- Press `2nd` and `Y=`. Select `Plot1...`. Press `ENTER` to turn **Plot1** `On`.

- Select `⊞` under `Type:` by pressing ⊙ once, `▶` five times, then `ENTER`.

- To ensure that the pairs of data will be plotted in the correct order, verify that **L1** is entered in `Xlist:` and **L2** is entered in `Freq:`.

  If they are not, press ⊙ once for `Xlist:`, `2nd` `LIST`, and `1`. Press ⊙ once more for `Freq:`, `2nd` `LIST`, and `2`.

- Press `WINDOW` so the histogram is set to the labels and quantities. To set the histogram with the range of our labels, in `Xmin=` press `1` `ENTER` and in `Xmax=` press `6` `ENTER`. Press `ENTER` twice more. To set the histogram with the range of our quantities, in `Ymin=` press `0` `ENTER` and in `Ymax=` press `4` `0` `ENTER`.

Step 3 Now graph the data using the **GRAPH** function.

- Press `GRAPH`.

- To make sure the labels and quantities are correctly graphed, press `TRACE` and `▶` or `◀`. The screen should display **L1:** and **L2:** at the bottom.

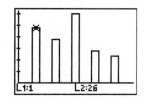

Sketch the graph in the space provided and label the axes, categories, and title.

Practice:

Use your graphing calculator to answer the following questions.

1. The most popular breeds of dog from a sample of animal lovers are listed in the table below.

Breed	Number
German Shepherd	42
Border Collie	84
Retriever	31
Terrier	34
Beagle	15

Which of the following best represents a bar graph of the data?

A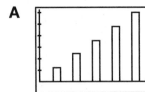

C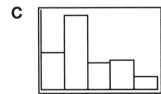

B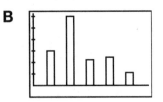

D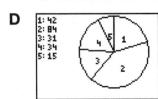

Saxon Math Course 3

2. The types of vehicles driven by the parents of students at a local middle school are listed in the table to the right. To collect the data, principals observed and recorded parent vehicles at morning drop-off.

Type of Vehicle	Number
Truck	206
Car	151
Compact Car	34
Van	7
Station wagon	3
Motorbike	1

a. Use your graphing calculator to generate a circle graph of the given data. Sketch the graph in the space provided, labeling the categories and percentages.

b. What are the most popular types of vehicles? Explain your answer.

c. The principals conclude that the data represents the types of vehicles driven by the parents for all 1100 students. Explain what could be wrong with that conclusion.

3. Joe decides to keep track of his expenses for a month. The totals for each category are listed in the table below.

Expense	Mortgage	Utilities	Food	Auto
Amount	$1,300	$300	$600	$500

    a. Choose a type of graph (bar graph, histogram, or circle graph) that does not show the quantity of each category. Explain your answer.

    b. Generate the chosen graph on the graphing calculator. Sketch the graph in the space provided, marking the appropriate labels.

    c. How could this graph be misleading to an advisor studying his finances?

• **Direct Variation**

Calculator Functions: Y= , TABLE , and WINDOW

Example: Which of the following is an example of direct variation? (The variables are underlined.)

A A taxi company charges three dollars to start the ride plus two <u>dollars</u> per <u>mile</u>.

B The <u>area</u> of a square is the square of the <u>length of its side</u>.

C The <u>perimeter</u> of a square is four times the <u>length of its side</u>.

Write down the function for each relationship.

A _____ **B** _____ **C** _____

Demonstration

Use the **Y=** and **TABLE** functions to find the example of direct variation.

Step 1 Enter the function $D = 2m + 3$ from example 1.

• Press Y= to go into **Y= Editor**.

TI-83+ Press 2 X,T,θ,*n* + . 3 .

TI-73 Press 2 *x* + 3 .

```
Plot1 Plot2 Plot3
\Y1 ⊟2X+3
\Y2=
\Y3=
\Y4=
\Y5=
\Y6=
\Y7=
```

Step 2 Generate a table from the function.

• Press 2nd and GRAPH .

```
 X  | Y1 |
 0  | 3  |
 1  | 5  |
 2  | 7  |
 3  | 9  |
 4  | 11 |
 5  | 13 |
 6  | 15 |
X=0
```

Fill in the table for the appropriate function and check for a constant ratio in space provided below.

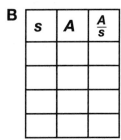

A

Miles	Dollars	Dollars/mile

B

s	A	A/s

C

s	P	P/s

Step 3 Graph $D = 2m + 3$ using the **WINDOW** function of the calculator.

- Set the graph with the domain and range that is appropriate for our scenario.

TI-83+ In ⌐Xmin=⌐ press ⓪ ⟨ENTER⟩, in ⌐Xmax=⌐ press ⑥ ⟨ENTER⟩, and in ⌐Xscl=⌐ press ① ⟨ENTER⟩. In ⌐Ymin=⌐ press ⓪ ⟨ENTER⟩, in ⌐Ymax=⌐ press ① ⓪ ⟨ENTER⟩, and in ⌐Yscl=⌐ press ① ⟨ENTER⟩.

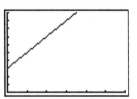

TI-73 In ⌐Xmin=⌐ press ⓪ ⟨ENTER⟩, in ⌐Xmax=⌐ press ⑥ then ⟨ENTER⟩ twice, and in ⌐Xscl=⌐ press ① ⟨ENTER⟩. In ⌐Ymin=⌐ press ⓪ ⟨ENTER⟩, in ⌐Ymax=⌐ press ① ⓪ ⟨ENTER⟩, and in ⌐Yscl=⌐ press ① ⟨ENTER⟩.

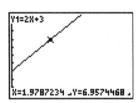

- Press ⟨GRAPH⟩ to see the function on the coordinate plane.

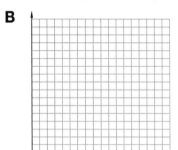

- To make sure the function matches its table, press ⟨TRACE⟩ and ⟨▶⟩ or ⟨◀⟩. The screen should display **X=** and **Y=** at the bottom.

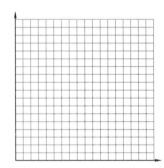

Graph the appropriate function and label the *x*- and *y*-axes and titles.

A   B   C

Repeat steps 1–3, making a table and sketching the graph for each of the remaining functions.

Name _____

Which function is an example of direct variation? Explain your answer.

Practice

Use your graphing calculator to answer the following questions.

1. Sunny's monthly cable <u>bill</u> is $43.95 plus $3.50 for each <u>movie</u> she orders. Write a function for Sunny's cable bill.

   a. Function: _____

   b. Make a table for the function, and check for a constant ratio in the space provided.

   c. Graph the function only with real-world values and label the *x*- and *y*-axes, and titles.

   d. Is the function an example of direct variation? Explain.

2. The amount of time Sully takes on a math test can be expressed with the function $t = 30p$, where t is the total time in seconds and p is the number of problems completed.

    a. Make a table for the function, and check for a constant ratio in the space provided.

    b. Is the function an example of direct variation? Explain.

    c. Using the **TABLE** function on the graphing calculator, how many minutes would it take Sully to complete a 50-problem math test?

        A 15 minutes

        B 25 minutes

        C 30 minutes

        D 1500 minutes

3. It takes each worker an hour to produce 3 computers on a production line. If there is a direct relationship between the number of workers and computers produced, how many workers would be needed to produce 42 computers in one hour?

    A 14   **B** 14.04   **C** 126   **D** 127.9

• Probability Simulation

New Calculator Function: MATH randInt(

Example: Read each of the following descriptors of events and explain how you would simulate each one. Then choose one of the events and conduct a simulation. Select a simulation tool for the experiment. Draw the chart you would use to organize your results.

1. A basketball player makes an average of 4 in 6 free throws. If he shoots 3 free throws per game, what is the probability he makes all of them?

2. John has 2 blue socks and 4 black socks in a drawer. On a dark morning, he reaches in and pulls out 2 socks. What is the probability that he has a matching pair?

3. One-half of the chairs at a party have stickers secretly placed under the seats entitling the person sitting in the chair to a door prize. What is the probability that Tina and her two friends will all sit in chairs with stickers under the seats?

Demonstration

Step 1 Assign integers to each possibility in all three events. For example, rainy weather could be assigned a 0, party cloudy could be a 1, and sunny skies could be a 2.

Event 1: _____

Event 2: _____

Event 3: _____

Step 2 Set up the **random integer** function in the graphing calculator for the event you choose. For example, to simulate the weather you could use randInt(0,2).

TI-83+   Press **MATH** and then (▶) three times. Select randInt(.
Enter the lowest assigned number, press (**,**), enter
the highest assigned number, and press (**)**).

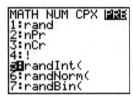 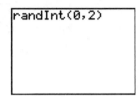

TI-73   Press **MATH** and then (▶) twice. Select randInt(. Enter
the lowest assigned number, press (**,**), enter the
highest assigned number, and press (**)**).

 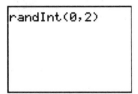

Step 3   Use the **random integer** function to simulate the event.

• Press **ENTER** at least 6 times.

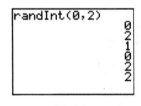

       TI-83+   *TI-73*

Draw a chart to organize the results.

Saxon Math Course 3

Compute the experimental probability of the event you chose. For example, the probability that it will rain is $\frac{2}{6}$ or $\frac{1}{3}$.

Practice

Use your graphing calculator to answer the following questions.

1. The odds of winning a particular raffle are 1 to 999. One ticket is drawn randomly from a basket. Danny has one ticket.

   a. Explain how you would simulate the drawing in the raffle using the graphing calculator. Choose one number to represent the ticket Danny holds.

   b. Use the **random integer** function to simulate the event 25 times. Draw a chart to organize the results.

   c. Why is it desirable to simulate the experiment rather than actually perform the experiment?

2. There are 14 students in a class, of which 4 are boys and 10 are girls. A student is randomly chosen to approach the board and solve a math problem.

    a. Using the given information, explain how you would simulate the event using a graphing calculator.

    b. Suppose there are 7 students in the class, of which 2 are boys and 5 are girls. Explain how you would simulate this event using a graphing calculator.

    c. Do you think the experimental probabilities in a. would be very different from b.? Explain.

Name _____

• Formulas for Sequences

New Calculator Functions: $\boxed{\text{seq(}}$

Example: Find the 8th and 9th terms of the sequence generated by the formula $a_n = 3n$.

___

Demonstration

Step 1 Find the existing terms of $a_n = 3n$ by using the **sequence** function.

• Go to the **LIST** menu.

TI-83+   Press $\boxed{\text{2nd}}$ $\boxed{\text{STAT}}$ to go into the **OPS** menu. Press $\boxed{\blacktriangleright}$ once and select $\boxed{\text{seq(}}$.

Press $\boxed{3}$ $\boxed{\text{X,T,Θ,}n}$ $\boxed{,}$ $\boxed{\text{X,T,Θ,}n}$ $\boxed{,}$ $\boxed{1}$ $\boxed{,}$ $\boxed{7}$ $\boxed{)}$ and $\boxed{\text{ENTER}}$. This function will calculate the expression $3x$ using x as the input variables from 1 to 7.

TI-73   Press $\boxed{\text{2nd}}$ $\boxed{\text{LIST}}$ to go into the **OPS** menu. Press $\boxed{\blacktriangleright}$ once and select $\boxed{\text{seq(}}$.

Press $\boxed{3}$ $\boxed{x}$ $\boxed{,}$ $\boxed{x}$ $\boxed{,}$ $\boxed{1}$ $\boxed{,}$ $\boxed{7}$ $\boxed{)}$. Press $\boxed{\text{ENTER}}$. This function will calculate the expression $3x$ using x as the input variable from 1 to 7.

• Press $\boxed{\blacktriangleright}$ repeatedly to see the 1st to 7th terms of the sequence.

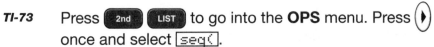

Find the first 7 terms of the sequence $a_n = 3n$.

Step 2 Find the terms of the sequence, including the 8th and 9th terms, using the **ENTRY** function.

• Press $\boxed{\text{2nd}}$ $\boxed{\text{ENTER}}$ to copy the **sequence** function. Press $\boxed{\blacktriangleleft}$ two times and then $\boxed{9}$.

- Press ENTER. This function displays the sequence from the 1st to the 9th term. Press (▶) repeatedly to see all the terms in the sequence.

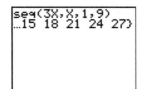

Find the first 9 terms of the sequence $a_n = 3n$.

Step 3 Find only the 8th and 9th terms of the sequence.

- Press 2nd ENTER to copy the **sequence** function. Press (◀) four times and then press 8.

- Press ENTER. This function displays the 8th and 9th terms of the sequence.

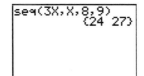

Find the 8th and 9th terms of the sequence $a_n = 3n$.

Practice

Use your graphing calculator to answer the following questions.

1. Find the following terms for the sequence $a_n = n(n + 1)$.

    a. Write down the graphing calculator sequence function you would use.

    b. Find the first 5 terms of the sequence.

    c. Find the 11th–15th terms of the sequence.

 Saxon Math Course 3

2. Below are the first 5 terms of a sequence.

$$3, 8, 15, 24, 35, \ldots$$

a. Use the **sequence** function in the graphing calculator to find which of the following expressions is correct.

A $3n$

B $2n - 1$

C $2n + 2$

D $n(n + 4) - 2n$

b. Find the 100th term of the sequence. _____

3. Which sequence can be used to find the nth term?

n	1	2	3	4
Term	$\frac{1}{2}$	$\frac{1}{4}$	$\frac{1}{6}$	$\frac{1}{8}$

A $a_n = \frac{1}{2}n$

B $a_n = \frac{1}{2n}$

C $a_n = \frac{1}{2^n}$

D Cannot be determined

• Scatterplots

New Calculator Functions: ⌊·⋅⌋ , ⌊⤳⌋ , and ⌊LinReg(ax+b)⌋

Example: Determine whether the wages of college students and the wages of high-school-only graduates listed below are correlated. If so, find the equation of a best-fit line.

Median Weekly Earnings, 2000

	High School Diploma ($)	Bachelor Degree or Higher ($)
Cashier	303	384
Cook	327	396
Computer Programmer	864	1039
Data-Entry Keyers	475	514
Designers	633	794
Electricians	714	976
Police	726	886
Legal Assistants	563	725
Real Estate Sales	614	918

Demonstration

Step 1 To graph the pairs of numbers from the table in the calculator, first enter them into **L1** and **L2**.

• Go to the **STAT** menu.

TI-83+ Press ⌊STAT⌋ and select ⌊Edit...⌋.

TI-73 Press ⌊LIST⌋.

• Enter the *x*-values from example 2 into **L1** by entering 303 and pressing ⌊ENTER⌋, and so on. Your screen should display **L1(10) =** after entering the last *x*-coordinate.

L1	L2	L3	1
475			
633			
714			
726			
563			
614			

L1(10) =

• Enter the *y*-values from example 2 into **L2** by pressing (▶) once. Enter 384 and press ⌊ENTER⌋, and so on. Your screen should display **L2(10) =** after entering the last *y*-coordinate.

L1	L2	L3	2
475	514		
633	794		
714	976		
726	886		
563	725		
614	918		

L2(10) =

 Saxon Math Course 3

Step 2 Once the pairs of numbers are entered, set the calculator to plot them on the coordinate plane.

- Press [2nd] and [Y=]. Select [Plot1...].

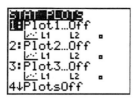

- Press [ENTER] to turn **Plot1** [On]. Select [..·] under [Type:] by pressing [⌄] once then [ENTER].

- To ensure that the pairs of data will be plotted in the correct order, verify that **L1** is entered in [Xlist:] and **L2** is entered in [Ylist:].

TI-83+   Press [⌄] once for [Xlist:] then [2nd] [1].
Press [⌄] once more for [Ylist:] then [2nd] [2].

TI-73   Press [⌄] once for [Xlist:] then [2nd] [LIST] [1]. Press [⌄] once more for [Ylist:] then [2nd] [LIST] [2].

Step 3 Now see what the pairs of numbers look like on a coordinate plane.

- Press [ZOOM] and select [ZoomStat].

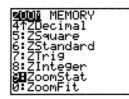

TI-83+   TI-73

- To make sure the pairs of numbers are correctly graphed, press [TRACE] and [▶] or [◀]. The screen should display **X=** and **Y=** at the bottom for each pair.

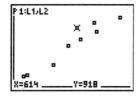

Sketch the scatterplot in the first quadrant of the coordinate plane in the space provided below. Be sure to label appropriately.

Step 4   To find a correlation between the data, follow the directions for the calculator you are using.

TI-83+   Sort the data into ordered pairs.

• Return to the **LIST** menu by pressing 2nd and STAT.

• Press (▶) once and select Sort A(from the **OPS** submenu.

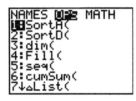

• Now press 2nd 1 , 2nd 2 and). This function will sort the data points in **L1** and **L2** as ascending ordered pairs. Press ENTER. The screen should display **Done**.

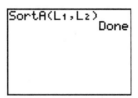

• To see the sorted data, return to the **Edit** submenu by pressing STAT and then selecting Edit.... If you press (▼) to scroll, you can see that the x-values increase in value and the y-values stay with their x-values.

TI-73 Sort the data into ordered pairs.

- Return to the **LIST** menu by pressing [2nd] and [LIST].

- Press (▶) once and select [SortA(] from the **OPS** submenu.

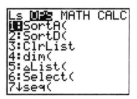

- Now press [2nd] [LIST] [1] [,] [2nd] [LIST] [2] and [)]. This function will sort the data points in **L1** and **L2** as ascending ordered pairs. Press [ENTER]. The screen should display **Done**.

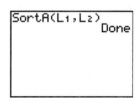

- To see the sorted data, return to the **LIST** menu by pressing [LIST]. If you press (▼) to scroll, you can see that the *x*-values increase in value and the *y*-values stay with their *x*-values.

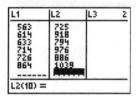

Step 5 Once the pairs of data are sorted from least to greatest, set the calculator to plot them as a line graph.

- Press [2nd] [Y=]. Select [Plot1...].

TI-83+ Select [⟋⟍] under [Type:] by pressing (▼) once, (▶) once, then [ENTER].

TI-73 Press [ENTER] to turn **Plot1** [On]. Select [⟋⟍] under [Type:] by pressing (▼) once, (▶) once, then [ENTER].

Step 6   Now see what the pairs of numbers look like as a line graph.

• Press [ZOOM] and select [ZoomStat].

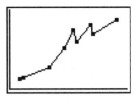

Are the wages of college graduates and the wages of
high-school-only graduates correlated? Explain your answer.

Step 7   If you determine that the pairs of numbers are correlated, find the
equation for the best-fit line.

• Calculate the best-fit line using the **regression** function.

TI-83+   Press [STAT] and ⊳ once to go to the **CALC** submenu.
Select [LinReg(ax+b)]. Press [ENTER]. The screen
should display a linear equation with a coefficient *a* and
constant *b*.

```
EDIT CALC TESTS
1:1-Var Stats
2:2-Var Stats
3:Med-Med
4:LinReg(ax+b)
5:QuadReg
6:CubicReg
7↓QuartReg
```
```
LinReg
 y=ax+b
 a=1.291547346
 b=-12.06506654
```

TI-73   Press [2nd] [LIST] and then ⊳ three times to go to
the **CALC** submenu. Select [LinReg(ax+b)]. Press
[ENTER]. The screen should display a linear equation with
a coefficient *a* and constant *b*.

```
Ls OPS MATH CALC
1:1-Var Stats
2:2-Var Stats
3:Manual-Fit
4:Med-Med
5:LinReg(ax+b)
6:QuadReg
7:ExpReg
```
```
LinReg
 y=ax+b
 a=1.291547346
 b=-12.06506654
```

Find the equation for the best-fit line. Round each value to the
nearest hundredth.

Saxon Math Course 3

Step 8 Graph the regression equation with the scatterplot of the data.

 • Press [2nd] [Y=] and select $\boxed{\texttt{Plot1...}}$. Select $\boxed{\cdot\cdot\cdot}$ under $\boxed{\texttt{Type:}}$ by pressing ⌄ once then (ENTER).

 • Go into **Y= Editor**. Press [Y=].

TI-83+   Press [1][·][2][9][X,T,θ,*n*] and [+] then [(−)][1][2][·][1].

TI-73   Press [1][·][2][9][*x*] and [+] then [(−)][1][2][·][1].

 • Press [ZOOM] and select $\boxed{\texttt{ZoomStat}}$.

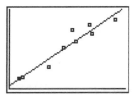

Sketch the scatterplot and the best-fit line on the first quadrant of the coordinate plane in the space provided below. Be sure to label appropriately.

Saxon Math Course 3   **113**

Practice

Use your graphing calculator to answer the following questions.

1. Find out how long it takes an increasing number of people to do the "wave." Separate into groups of 9 or fewer students. Each group should have a stop watch and sit down in one row. From the left end of the row, time one student standing up, raising the arms, making a shout, lowering the arms, and then sitting back down. Record the data in the table. Now record the amount of time it takes two students to do the wave. The second student should start as soon as the student to his/her left sits back down. Record this data in the table. Continue timing the wave with an increasing number of students and recording the data in the table until the last wave is done with all the students in the group.

Number of People									
Time (in seconds)									

a. Sketch the scatterplot in the first quadrant of the coordinate plane in the space provided below. Be sure to label appropriately.

b. Is there a correlation between the number of people and the time it takes to do the wave? Explain your answer.

c. Find the equation for the best-fit line. Round each value to the nearest hundredth.

2. The table to the right lists the ages and the amount of time (in hours) that each child sleeps.

   a. Are age and hours of sleep correlated? Explain your answer.

Age (years)	Sleep (hours)
10	10
8	10.5
11	9
12	8
10	11
9	10
13	7.5
15	7
16	7

b. Find the equation for the best-fit line. Round each value to the nearest hundredth.

c. Sketch the scatterplot and the best-fit line on the first quadrant of the coordinate plane in the space provided below. Be sure to label appropriately.

• Graphing Equations Using Intercepts

Calculator Functions: STAT and Y=

Example: Find the *x*- and *y*-intercepts of $3x + 4y = 24$. Then graph the equation.

The *y*-intercept is _____. The *x*-intercept is _____.

Demonstration

Step 1 To graph the intercepts in the calculator, first enter them into **L1** and **L2**.

   • Go to the **STAT** menu.

TI-83+ Press STAT and select Edit....

TI-73 Press LIST.

   • Enter the *x*-values from example 1 into **L1** by pressing
   0 ENTER then 8 ENTER. Your screen should
   display **L1(3) =** after entering the last *x*-coordinate.

   • Enter the *y*-values from example 1 into **L2**
   by pressing ▶ once. Then press
   6 ENTER and 0 ENTER.Your screen
   should display **L2(3) =** after entering the
   last *y*-coordinate.

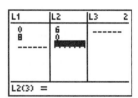

Step 2 Once the pairs of numbers are entered, set the calculator to plot them on
the coordinate plane.

   • Press 2nd and Y= . Select Plot1....

   • Press ENTER to turn **Plot1** On. Select |·˙· under Type:
   by pressing ⌄ once then ENTER .

   • To ensure that the pairs of data will be plotted in the
   correct order, verify that **L1** is entered in Xlist: and **L2**
   is entered in Ylist:.

TI-83+ Press ⌄ once for Xlist:, 2nd , and 1 . Press
   ⌄ once more for Ylist:, 2nd , and 2 .

     Saxon Math Course 3

TI-73   Press ⌄ once for `Xlist:`, then [2nd] [LIST], and
[1]. Press ⌄ once more for `Ylist:`, [2nd],
[LIST], and [2].

Transform $3x + 4y = 24$ to the slope-intercept form.

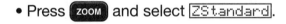

Step 3   Graph the linear equation with the intercepts.

- Go into **Y= Editor**. Press [Y=].

TI-83+   Press [(] [(−)] [3] [÷] [4] [)] [X,T,θ,n]
and [+] then [6].

TI-73   Press [(] [(−)] [3] [÷] [4] [)] [x]
and [+] then [6].

- Press [ZOOM] and select `ZStandard`.

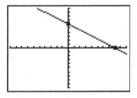

Sketch the intercepts and the line on a coordinate plane in the
space provided below. Be sure to label the graph appropriately.

Name _____

_____

Practice

Use your graphing calculator to answer these questions.

1. Find the *x*- and *y*-intercepts of $2x - 5y = 30$. Then graph the equation.

a. What are the intercepts?

The *y*-intercept is _____. The *x*-intercept is _____.

b. Transform $2x - 5y = 30$ to the slope-intercept form.

c. Sketch the intercepts and the line on a coordinate plane in the space provided below. Be sure to label the graph appropriately.

 Saxon Math Course 3

2. Consider the equation $-1.5x + y = 30$.

    a. What are the intercepts to the equation?

        The *y*-intercept is _____. The *x*-intercept is _____.

    b. Transform $-1.5x + y = 30$ to the slope-intercept form.

    c. Sketch the intercepts and the line on a coordinate plane in the space provided below. Be sure to label the graph appropriately.

• **Solving Problems with Two Unknowns by Graphing**

Calculator Functions: Y= , TRACE , and TABLE

Example: Theo is thinking of two numbers. He says that the sum of the numbers is 6. He also says that one number is 10 more than the other number. What are the two numbers? Solve this system of equations by graphing.

$$\begin{cases} x + y = 6 \\ y = x + 10 \end{cases}$$

Transform $x + y = 6$ and $y = x + 10$ to the slope-intercept form.

Demonstration

Step 1 To solve the system from the example, first enter both equations into the calculator.

    • Go into **Y= Editor**. Enter the first equation into **Y1**.

TI-83+   Press Y= . Press (−) X,T,θ,n + 6 .

TI-73   Press Y= . Press (−) x + 6 .

    • Enter the second equation into **Y2**.

TI-83+   Press ENTER . Press X,T,θ,n + 1 0 .

TI-73   Press ENTER . Press x + 1 0 .

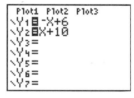

Step 2 Graph the system by using the **ZOOM** function on the calculator.

    • Press ZOOM and select ZStandard .

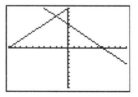

Saxon Math Course 3

Graph the system of equations on a coordinate plane in the space provided.

Step 3 Find the solution to the system by finding the point of intersection.

- Press [TRACE] and (◀) or (▶). Place the cursor near the point of intersection from the graph. The function gives us an approximation for the solution.

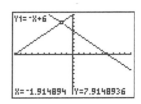

- Use the approximation from the graph to find the exact solution. Press [2nd] [GRAPH] and (▲) or (▼). Place the cursor near the same x-value of the point of intersection as you found from the graph. The actual point of intersection will have the same y-value for both **Y1** and **Y2**.

The solution to the system of equations is _____.

Practice

Use your graphing calculator to answer the following questions.

1. Solve the system of equations by graphing.

$$\begin{cases} 2x + y = -2 \\ -2x + y = 6 \end{cases}$$

    a. Transform each equation to the slope-intercept form.

    b. Graph the system of equations on a coordinate plane in the space provided.

    c. The solution to the system of equations is _____ .

Saxon Math Course 3

2. A music band has decided to go on tour. It expects to sell each ticket at an average of $15, and each theatre will pay them $125. The stage fee is $305, and equipment rental will cost $5 per ticket. Let x be the number of tickets sold and y be the ticket income or show costs.

$$\begin{cases} -15x + y = 125 \\ -305 + y = 5x \end{cases}$$

a. Transform each equation to the slope-intercept form.

b. Graph the system of equations on a coordinate plane in the space provided.

c. How many tickets must the band sell to cover the costs of each show of the tour?

• Inverse Variation

Calculator Functions: Y= and TABLE

Example: Robots are programmed to assemble widgets on an assembly line. The number of robots working (*n*) and the amount of time (*t*) it takes to assemble a fixed number of widgets are inversely proportional. If it takes 10 robots 8 hours to assemble a truckload of widgets, how long would it take 20 robots to assemble the same number of widgets?

Write an equation that describes the number of robots working (*n*) and the amount of time (*t*) relationship.

Transform the equation to solve for *t*.

Demonstration

Step 1 Enter the inverse equation from example 2 into the calculator.

• Press Y= to go into **Y= Editor**.

TI-83+ Press 8 0 ÷ X,T,Θ,*n* .

TI-73 Press 8 0 ÷ *x* .

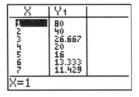

Step 2 Generate the table from the inverse equation.

• Press 2nd and GRAPH .

Saxon Math Course 3

Name _____

• Use ⌄ or ⌃ to find the pairs of values in whole numbers.

X	Y₁	
74	1.0811	
75	1.0667	
76	1.0526	
77	1.039	
78	1.0256	
79	1.0127	
80	1	
X=80		

Make a table for the inverse equation in the space provided below using only the whole numbers.

Step 3   Graph the inverse equation using the **WINDOW** function of the calculator.

• Set the graph with the domain and range that is appropriate for our scenario.

TI-83+   In $\boxed{Xmin=}$ press $\boxed{(-)}$ $\boxed{5}$ $\boxed{ENTER}$, in $\boxed{Xmax=}$ press $\boxed{5}$ $\boxed{0}$ $\boxed{ENTER}$, and in $\boxed{Xscl=}$ press $\boxed{1}$ $\boxed{ENTER}$. In $\boxed{Ymin=}$ press $\boxed{(-)}$ $\boxed{5}$ $\boxed{ENTER}$, in $\boxed{Ymax=}$ press $\boxed{5}$ $\boxed{0}$ $\boxed{ENTER}$, and in $\boxed{Yscl=}$ press $\boxed{1}$ $\boxed{ENTER}$.

```
WINDOW
 Xmin=-5
 Xmax=50
 Xscl=1
 Ymin=-5
 Ymax=50
 Yscl=1
 Xres=1
```

TI-73   In $\boxed{Xmin=}$ press $\boxed{(-)}$ $\boxed{5}$ $\boxed{ENTER}$, in $\boxed{Xmax=}$ press $\boxed{5}$ $\boxed{0}$ and then $\boxed{ENTER}$ twice, and in $\boxed{Xscl=}$ press $\boxed{1}$ $\boxed{ENTER}$. In $\boxed{Ymin=}$ press $\boxed{(-)}$ $\boxed{5}$ $\boxed{ENTER}$, in $\boxed{Ymax=}$ press $\boxed{5}$ $\boxed{0}$ $\boxed{ENTER}$, and in $\boxed{Yscl=}$ press $\boxed{1}$ $\boxed{ENTER}$.

```
WINDOW
 Xmin=-5
 Xmax=50
 ΔX=.5851063829…
 Xscl=1
 Ymin=-5
 Ymax=50
 Yscl=1
```

• Press $\boxed{GRAPH}$ to see the inverse equation on the coordinate plane.

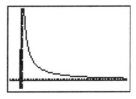

Graph the inverse equation, and label appropriately.

How long would it take 20 robots to assemble a truckload of widgets?

Practice

Use your graphing calculator to answer these questions.

1. The number of students working on a science fair project and the number of hours it takes to complete it are inversely related. With a team of 2 it takes 30 hours to complete the project.

   a. Write an equation that describes the number of students (s) and the number of hours (t) relationship.

   b. Transform the equation to solve for t. _____

   c. Make a table for the inverse equation up to 6 students in the space provided below.

d. Graph the inverse equation, marking the appropriate labels.

e. How long would it take 5 people to complete the project?

2. The number of 15-minute breaks a production employee receives in an 8-hour day is inversely related to the number of defects he/she commits. Working with 4 breaks averages 14 defects. This relationship can be expressed as the equation $b \cdot d = 56$, where b is the number of breaks and d is the number of defects.

   a. Transform the equation to solve for d. _____

   b. Make a table for the inverse equation for every 2 breaks, starting at 2 and ending at 10, in the space provided below. Round to the nearest whole number.

c. If you were an employer, would it be an advantage to give an employee 8 breaks versus 6 breaks? Explain.

Name _____

• Box-and-Whisker Plots

Calculator Functions: STAT , PLOT , and TRACE

Example: During physical fitness testing, students count the number of curl-ups they can do in one minute. The following totals were collected for the class.

$$32, 45, 36, 40, 29, 50, 39, 45, 52$$
$$26, 38, 48, 55, 40, 38, 51, 35, 40$$
$$53, 42, 39, 46, 43, 40, 38, 45, 41$$

Create a box-and-whisker plot for the data in example 1.

Demonstration

Step 1 Enter the data into the calculator from example 1.

• Go to the **STAT** menu.

TI-83+ Press STAT and select Edit....

TI-73 Press LIST .

• Enter the totals from example 1 into **L1** by entering 32 and pressing ENTER , entering 45 and pressing ENTER , and so on. Your screen should display **L1(28) =** after entering the last data point.

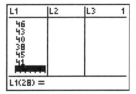

Step 2 Once the data are entered, set the calculator to display them in a box-and-whisker plot.

• Press 2nd and Y= . Select Plot1....

• Press ENTER to turn **Plot1** On.

• Select ⊞ under Type:.

TI-83+ Press ▼ once, ▶ four times, then ENTER .

TI-73 Press ▼ once, ▶ six times, then ENTER .

- To ensure that the correct data will be plotted, verify that
  L1 is entered in ⟨Xlist:⟩ and ⟨Freq:⟩ is set to **1**.

TI-83+   Press ⟨▼⟩ once for ⟨Xlist:⟩, ⟨2nd⟩, and ⟨1⟩. Press
⟨▼⟩ once more for ⟨Freq:⟩, ⟨ALPHA⟩, and ⟨1⟩.

TI-73   Press ⟨▼⟩ once for ⟨Xlist:⟩, ⟨2nd⟩, ⟨LIST⟩, and ⟨1⟩.
Press ⟨▼⟩ once more for ⟨Freq:⟩ and ⟨1⟩.

Step 3   Now see what the data look like as a box-and-whisker plot.

- Press ⟨ZOOM⟩ and select ⟨ZoomStat⟩.

TI-83+   TI-73

- To find the values of each quartile, press ⟨TRACE⟩ and ⟨▶⟩
  or ⟨◀⟩. The screen should display **minX=26,
  Q1=38, Med=40, Q3=46 and maxX=55**.

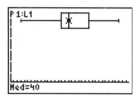

Draw the box-and-whisker plot of the data above a number
line. Label the values for each quartile. Use the space provided
below.

Saxon Math Course 3

___

Practice

Use your graphing calculator to answer the following questions.

1. Roll two number cubes 25 times and record the sum of the faces showing up. Enter your data into **L1** on the calculator. Then use the **Plot** function to draw a box-and-whisker plot.

    a. Draw the box-and-whisker plot of the data above a number line. Label the values for each quartile. Use the space provided below.

    b. What is the median of the data? _____

    c. What is the median of the first half of the data? _____

    d. What is the median of the second half of the data? _____

2. Twenty middle school students were surveyed about the number of hours they spend studying each week. The results are listed below.

35	40	25	17	11	5	1	0	27	32
24	17	16	12	12	9	5	6	8	14

    a. Draw a box-and-whisker plot of the data above a number line. Label the values for each quartile. Use the space provided below.

b. According to the box-and-whisker plot, how much time does a typical middle school student spend studying during a given week? Explain your answer.

c. Find the lower and upper extremes. Which extreme is less like the majority of the data? Explain your answer.

3. Below are the weight gains (in ounces) for nine baby chicks over an 8-week period.

3.9	4.1	4.5	4.8	4.9	4.4	4.6	5.0	4.4

What is the interquartile range of the data?

A 0.6 ounces

B 1.1 ounces

C 4.5 ounces

D 5.0 ounces

• Consumer Interest

New TI-83+ Calculator Function: TVM Solver...

Example: Jason has a credit card balance of $1200. The card company charges 2% interest on the unpaid balance each month.

    a. What is the interest charge on Jason's current balance?

    b. Jason has a choice of paying the balance plus interest in full, or making a minimum payment of $50. If Jason makes only the minimum payment each month, what balance will he carry to the next month?

    c. If Jason makes only the minimum payment and makes an additional $800 in purchases during the next month, what will be Jason's approximate balance after the next month's interest is applied?

Demonstration

This demonstration is for the **TI-83+** graphing calculator.

    Step 1   Calculate the interest charged using the **TVM Solver** from example a.

        • Press APPS and select Finance... . Select TVM Solver... in the **CALC** submenu.

  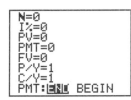

        • Set the function to solve for the interest charged after one month by pressing 1 for **N=** then ENTER , 2 4 for **I%** then ENTER , 1 2 0 0 for **PV=** then ENTER , and 0 for **PMT=** then ENTER .

        • Check that **12** is entered into **P/Y=** and **C/Y=** and that **END** is highlighted black. If not, press ⌄ once from **PMT=** and enter 12, press ⌄ once more and enter 12, then press ⌄ and ENTER .

• To **SOLVE** for the balance with interest, use 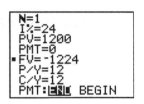 or ⌄ to go to **FV=**. Then press ALPHA ENTER. The screen should display **–1224**.

```
N=1
I%=24
PV=1200
PMT=0
■FV=-1224
P/Y=12
C/Y=12
PMT:END BEGIN
```

What is the interest charge on Jason's current balance?

Step 2   Calculate the balance with a payment using the **TVM Solver** from example b.

• Use ⌃ or ⌄ to go to **PMT=**. Press (–) 5 0 ENTER.

• Check that the same values from example a. are entered into the other functions. If not, use ⌃ or ⌄ to get to **N=**. Press 1 then ENTER, 2 4 for **I%** then ENTER, and 1 2 0 0 for **PV=** then ENTER.

```
N=1
I%=24
PV=1200
PMT=-50
FV=0
P/Y=12
C/Y=12
PMT:END BEGIN
```

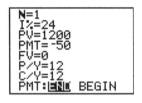

• To **SOLVE** for the balance with a minimum payment, use ⌃ or ⌄ to go to **FV=**. Then press ALPHA and ENTER. The screen should display **–1174**.

```
N=1
I%=24
PV=1200
PMT=-50
■FV=-1174
P/Y=12
C/Y=12
PMT:END BEGIN
```

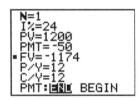

If Jason makes only the minimum payment, what balance will he carry to the next month?

Step 3   Calculate the balance with additional purchases using the **TVM Solver** from example c.

Assume that the purchases were made on the last day of the billing cycle. Find next month's balance without the additional purchases.

• Use ⌃ or ⌄ to go to **PV=**. Enter 1174 then press ENTER. Use ⌃ or ⌄ to go to **PMT=**. Press 0 then ENTER.

```
N=1
I%=24
PV=1174
PMT=0
FV=0
P/Y=12
C/Y=12
PMT:END BEGIN
```

- To **SOLVE** for the balance without additional purchases, use ⊙ or ⊙ to go to **FV=**. Then press (ALPHA) and (ENTER). The screen should display **−1197.48**.

Assume that the purchases were made on the first day of the billing cycle. Find next month's balance with the additional purchases.

- Use ⊙ or ⊙ to go to **PV=**. Enter 1174, press (+), enter 800, then press (ENTER). The screen should display **1974**.

- To **SOLVE** for the balance with additional purchases, use ⊙ or ⊙ to go to **FV=**. Then press (ALPHA) and (ENTER). The screen should display **−2013.48**.

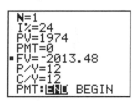

What will be Jason's approximate balance after the next month's interest is applied?

Practice

Use your graphing calculator to answer the following questions.

1. Ryan wants to go on a vacation to Hawaii for $5400. He will charge it on his credit card, which has a 2.499% monthly interest rate. What will his balance be one month after the vacation?

   a. List the values you enter into the **TVM Solver** before finding the answer.

   N= _____   FV= _____

   I%= _____   P/Y= _____

   PV= _____   C/Y= _____

   PMT= _____   PMT: _____

   b. SOLVE for FV=. _____

   c. How much interest will he owe one month after the vacation? Round to the nearest cent. _____

2. Sal is looking to buy a car at a sticker price of $17,500. He would like to take out a loan that has a 0.17% monthly rate and payments of $500 per month. What will be his balance after one year of the loan?

   a. List the values you enter into the **TVM Solver** before finding the answer.

     N= _____   FV= _____

     I%= _____   P/Y= _____

     PV= _____   C/Y= _____

     PMT= _____   PMT: _____

   b. SOLVE for **FV=**. _____

   c. In real-world terms, what is Sal's balance after one year? Round to the nearest cent.

     A −$23,916.78

     B −$11,803.94

     C $11,803.94

     D $23,916.78

3. Ms. Wren wants to send her daughter to college. Tuition will cost $13,500, and books will cost $9000 for four years at State U. If Mrs. Wren borrows the money from the bank at an annual rate of 6.7% and makes $350 in monthly payments, what will be the loan's balance after her daughter finishes college?

   a. List the values you enter into the **TVM Solver** before finding the answer.

     N= _____   FV= _____

     I%= _____   P/Y= _____

     PV= _____   C/Y= _____

     PMT= _____   PMT: _____

   b. SOLVE for **FV=**. _____

c. How much money will Ms. Wren still owe after her daughter's four years of college? Round to the nearest cent. _____

d. What could she do to lower her remaining balance?

Name _____

• Non-Linear Functions

Calculator Functions: [**Y=**], [**WINDOW**], and [**TRACE**]

Example: Leslie held a softball 5 ft above the ground and threw it straight up at 32 ft/sec. She has studied physics and predicted that, considering only gravity, the height (*h*) in feet of the ball at (*t*) seconds could be described by this function:

$$h = -16t^2 + 32t + 5$$

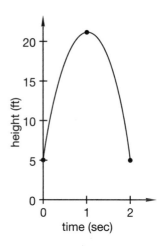

Demonstration

Step 1 To generate a table for the second activity from the function $h = -16t^2 + 32t + 5$, first enter it into the calculator.

• Go into **Y= Editor**.

TI-83+ Press [**Y=**]. Press [**(−)**] [**1**] [**6**] [**X,T,θ,n**] [**x²**] and [**+**] and [**3**] [**2**] [**X,T,θ,n**] and [**+**] then [**5**].

```
Plot1  Plot2  Plot3
\Y1◘-16X²+32X+5
\Y2=
\Y3=
\Y4=
\Y5=
\Y6=
\Y7=
```

TI-73 Press [**Y=**]. Press [**(−)**] [**1**] [**6**] [*x*] [*x²*] and [**+**] and [**3**] [**2**] [*x*] and [**+**] then [**5**].

```
Plot1  Plot2  Plot3
\Y1◘-16X²+32X+5
\Y2=
\Y3=
\Y4=
```

• To see the table of values for the function, press [**2nd**] and [**GRAPH**].

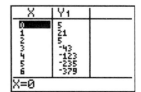

   Saxon Math Course 3

Complete the table to find the height of the ball at 0, 1, and 2 seconds.

t	0	1	2
h			

Step 2 Graph $h = -16t^2 + 32t + 5$ using the **WINDOW** function of the calculator.

• Set the graph with appropriate x- and y-values. Press
  WINDOW

TI-83+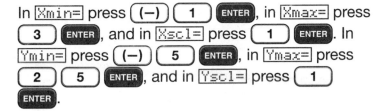
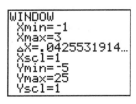

TI-73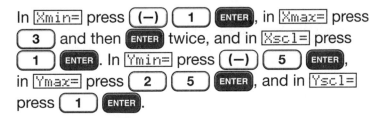
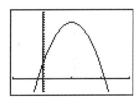

• Press **GRAPH** to see the function on the coordinate plane.

Sketch the graph for the function $h = -16t^2 + 32t + 5$, in the space provided below. Label the x- and y-axes, and the title.

Step 3 Find the maximum and minimum *y*-values using the **TRACE** function.

- Press TRACE, then use (◀) and (▶) to look for the **X=** when the **Y=** is at its greatest value.

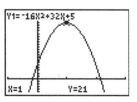

- Use (◀) and (▶) to find a positive value for **X=** when the **Y=** is just below 0.

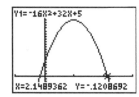

What is the maximum height of the ball? _____

When does it reach this height? _____

Approximately how long after it is thrown does the ball hit the ground?

Practice

Use your graphing calculator to answer the following questions.

1. A model rocket is launched 2 feet from the ground. It initially takes off at a speed of 100 ft/sec. Porcia figures out that the height (*h*) in feet of the rocket in seconds (*t*) could be determined by the function, $h = -16t^2 + 100t + 2$.

   a. Make a table for the heights of the rocket from 0 seconds to the time it hits the ground.

b. Sketch the graph for the function in the space provided below. Label the x- and y-axes and the title.

c. Approximately how high does the rocket go? _____

d. About how many seconds pass before the rocket hits the ground? _____

2. A particular summer concert promoter has found that attendance during the day is quadratic. Reyna has developed the equation $y = -0.5x^2 + 2.5x + 3.5$ to show the relationship, where x is the hour after 6 p.m. and y is the number of people at the festival (in thousands).

   a. Make a table for attendance from 6 p.m. to 12 midnight (hours 0–6).

   b. At what time does the concert have the greatest number of people? For what reason would there be a maximum number in attendance?

c. At what time does the concert have the fewest people? For what reason would there be a minimum number in attendance?

3. Kai throws a ball upward from the top of the bleachers for a science experiment. The height of the bleachers is 25 feet from the ground. Kai throws the ball at a average speed of 50 ft/sec, which is affected by the rate of gravitational pull, −16ft/sec^2. Let t be time in seconds and h be the height of the ball.

a. Which of the following functions correctly represents the path of Kai's ball?

A $h = -16t^2 + 75t$

B $h = -16t^2 + 50t + 25$

C $h = -16t^2 + 25t + 50$

D $h = 50t^2 + 25t - 16$

b. Sketch the graph of the function in the space provided below. Label the x- and y-axes, and the title.

c. Approximately, what is the maximum height of the ball? _____

How long will it take to reach its maximum height? _____

• Significant Digits

New Calculator Functions: [QuadReg]

Example: Pairs of students did an experiment to test memory. They used the random generator in the graphing calculator to list an increasing quantity of numbers. One student called out the list and another student repeated as many numbers as he/she could recall. The results from one pair's experiment are listed in the table.

Quantity Called Out	Quantity Recalled	Quantity Called Out	Quantity Recalled
1	1	11	10
2	2	12	10
3	3	13	9
4	4	14	8
5	5	15	8
6	6	16	6
7	7	17	5
8	8	18	1
9	9	19	2
10	7	20	0

Demonstration

Step 1 Graph the pairs of numbers from the table in the calculator by entering them into **L1** and **L2**.

• Go to the **STAT** menu.

TI-83+ Press (STAT) and select [Edit...].

TI-73 Press (LIST).

• Enter the *x*-values into **L1** by entering 1 and pressing (ENTER), and so on. Your screen should display **L1(21) =** after entering the last *x* value.

• Enter the *y*-values into **L2** by pressing (▶) once. Enter 1 and press (ENTER), and so on. Your screen should display **L2(21) =** after entering the last *y* value.

Step 2 Set the calculator to plot the pairs of numbers on the coordinate plane.

- Press [2nd] and [Y=]. Select [Plot1...].

- Press [ENTER] to turn **Plot1** [On]. Select [.·˙] under [Type:] by pressing [▼] once then [ENTER].

- To ensure that the pairs of data will be plotted in the correct order, verify that **L1** is entered in [Xlist:] and **L2** is entered in [Ylist:].

TI-83+ Press [▼] once for [Xlist:] then [2nd] [1].
Press [▼] once more for [Ylist:] then [2nd] [2].

TI-73 Press [▼] once for [Xlist:] then [2nd] [LIST] [1]. Press [▼] once more for [Ylist:] then [2nd] [LIST] [2].

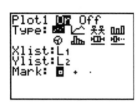

Step 3 Now see what the pairs of numbers look like on a coordinate plane.

- Press [ZOOM] and select [ZoomStat].

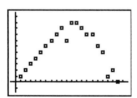

What is the shape of the scatterplot: linear or quadratic?
Explain your answer.

   Saxon Math Course 3

Step 4   Find the equation of best-fit.

• Calculate the best-fit equation using the **regression** function.

TI-83+   Press  and (▶) once to go to the **CALC** submenu. Select ⌐QuadReg¬. Press (ENTER).

TI-73   Press (2nd) (LIST) and then (▶) three times to go to the **CALC** submenu. Select ⌐QuadReg¬. Press (ENTER).

Find the equation of best-fit. Round to the first significant digit.

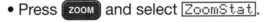

Step 5   Graph the regression equation with the scatterplot of the data.

• Go into **Y= Editor**. Press (Y=).

TI-83+   Press ((−)) (0) (.) (1) (X,T,θ,n) (x²) and (+) and (2) (X,T,θ,n) and (+) then ((−)) (2).

TI-73   Press ((−)) (0) (.) (1) (x) (x²) and (+) and (2) (x) and (+) then ((−)) (2).

• Press (ZOOM) and select ⌐ZoomStat¬.

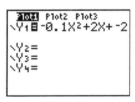

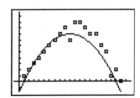

Name _____

Sketch the scatterplot and the best-fit equation in the first quadrant of the coordinate plane in the space provided below. Be sure to label appropriately.

Is the quadratic regression the equation of best-fit? Explain.

Recalculate the quadratic equation using step 4. Find the equation of best-fit, rounding to the fifth significant digit.

Step 6   Graph the new regression equation.

• Go into **Y= Editor**. Press  .

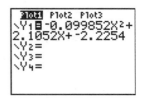

TI-83+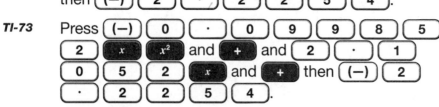

TI-73   Press (−) 0 · 0 9 9 8 5
2 x x^2 and + and 2 · 1
0 5 2 x and + then (−) 2
· 2 2 5 4 .

• Press **ZOOM** and select Z̲o̲o̲m̲S̲t̲a̲t̲ .

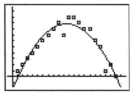

     Saxon Math Course 3

Sketch the scatterplot and the regression on the first quadrant of the coordinate plane in the space provided below. Be sure to label appropriately.

Is the recalculated regression a better equation of best-fit? Explain.

Practice

Use your graphing calculator to answer the following questions.

1. The following data set shows the various speeds (mph) of a car and the gas mileage (mpg) achieved.

Speed	15	23	30	35	42	45	50	54	60	65
mpg	14	17	20	24	26	23	18	15	11	10

   a. What is the shape of the scatterplot: linear or quadratic? Explain your answer.

   b. Find the equation of best-fit. Round to the second significant digit.

c. Sketch the scatterplot and the best-fit equation in the first quadrant of the coordinate plane in the space provided below. Be sure to label appropriately.

d. Is the regression the equation of best-fit? Explain.

• Sine, Cosine, Tangent

New Calculator Functions: [SIN], [COS], and [TAN]

Example: Use a calculator to find the sine and cosine of ∠A.

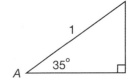

Demonstration

Step 1 Find the sine of ∠A.

    • Set the calculator to **Degree**.

TI-83+   Press [MODE]. Press ⌄ two times to go to
      radian. Press (▶) once then [ENTER] so that Degree
      is highlighted black.

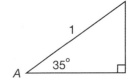

TI-73   Press [MODE]. Press ⌄ two times then [ENTER] so that
      Degree is highlighted black.

    Make sure all of the other first choices on each row are
    highlighted black. If not, use ⌃ or ⌄ and [ENTER] to
    set them. Then press [2nd] and [MODE] to quit.

    • Calculate the sine of ∠A.

TI-83+   Press [2nd] and [MODE]. Then press [SIN], enter 35,
      and press [ )] then [ENTER]. The screen should
      display a long decimal number.

TI-73   Press 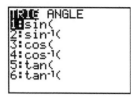 and to go to the **TRIG** submenu.
Select $\underline{sin(}$. Enter 35 and press then . The
screen should display a long decimal number.

What is the sine of $\angle A$ rounded to the nearest thousandth?

Step 2   Find the cosine of $\angle A$.

• Calculate the cosine of $\angle A$.

TI-83+   Press , enter 35, and press then .
The screen should display a long decimal number.

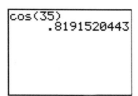

TI-73   Go back to the **TRIG** submenu by pressing
and . Select $\underline{cos(}$. Enter 35 and press
then . The screen should display a long decimal
number.

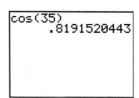

What is the cosine of $\angle A$ rounded to the nearest thousandth?

Draw and label the side lengths of the right triangle with
m∠A = 35° and a hypotenuse of 1. Use the space provided
below.

Example: Find the height of a tree if the angle of
elevation from 100 ft away is 22°.

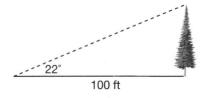

100 ft

Step 3 Find the tangent of the angle of elevation.

• Calculate the tangent of 22°.

TI-83+ Press [TAN], enter 22, and press [)] then [ENTER].
The screen should display a long decimal number.

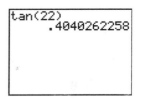

TI-73 Go back to the **TRIG** submenu by pressing [2nd]
and [DRAW]. Select [tan(]. Enter 22 and press [)]
then [ENTER]. The screen should display a long decimal
number.

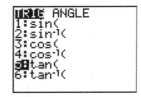 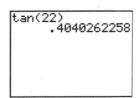

What is the tangent of 22° rounded to the nearest thousandth?

Use a proportion to find the height of the tree. _____

Practice

Use your graphing calculator to answer the following questions.

1. Find the sine, cosine, and tangent of a right triangle with a hypotenuse of 1 and angle of elevation of 45°.

   a. What is the sine of 45° rounded to the nearest thousandth?   _____

   b. What is the cosine of 45° rounded to the nearest thousandth?   _____

   c. What is the tangent of 45° rounded to the nearest thousandth?   _____

   d. What is special about the sine and cosine of 45° in a right triangle?

   e. What is special about the tangent of 45° in a right triangle?

2. A 21-foot tree needs trimming. The angle made by the ladder and the ground is 70°. Use sine to find the length of the ladder.

   a. What is the sine of 70° rounded to the nearest thousandth?

   b. Use a proportion to find the length of the ladder. Round to the nearest foot.

   c. Could you use tangent to find the length? Explain.

Saxon Math Course 3

3. A water skier must be at least a horizontal distance of 50 feet from the boat in order to safely avoid undertow from the propeller. If the angle of elevation is 35° from the skier to the pole, approximately how long must the rope be?

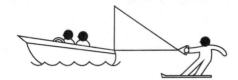

A 41 ft

B 61 ft

C 71 ft

D 87 ft

Calculations

Calculator Functions: [÷] [+] [ENTER] or [⅓]

- Enter the mixed numbers into the calculator.

TI-83+ Press [( ] whole number [+] numerator [÷] denominator [ )].

TI-73 Press [( ] whole number [+] [⅓] numerator [⌄] denominator [▶] [ )].

Calculator Functions: [MATH] [Frac]

- Convert a decimal to a fraction.

TI-83+ Subtract the integer part (if any) by pressing [−], entering the integer, then pressing [ENTER]. Then press [MATH], select [Frac], and press [ENTER].

Calculator Functions: [x^2] and [$\sqrt{\ }$]

- To take the square root of a number, press [2nd] and [x^2]. Enter the number and press [ )] [ENTER].

- To square a number, enter the number and press [x^2] [ENTER].

Calculator Functions: [^] and [$\sqrt[3]{(}$]

- To take the cubed root of a number, press [MATH] and select [$\sqrt[3]{(}$]. Enter the number and press [ )] [ENTER].

- To cube a number, enter the number and press [^] [3] [ENTER].

Calculator Functions: [STAT] and [LIST]

- Go to the **STAT** menu.

TI-83+ Press [STAT] and select [EDIT...].

TI-73 Press [LIST].

- Enter the data into into **L1** by entering the first number and pressing [ENTER], entering the second number and pressing [ENTER], and so on.

Saxon Math Course 3

Calculator Function: MATH

- Go to the **LIST** menu.

TI-83+   Press [2nd] and [STAT].

TI-73   Press [2nd] and [LIST].

- Press (▶) twice and select mean(or median(from the **MATH** submenu.

TI-83+   Press [2nd], (1) and ( )). Press [ENTER].

TI-73   Press [2nd] [LIST], select L1 and press ( )). Press [ENTER].

Calculator Functions: LinReg(ax+b) or QuadReg

- Calculate the best-fit equation.

TI-83+   Press [STAT] and (▶) once to go to the **CALC** submenu. Select LinReg(ax+b) or QuadReg. Press [ENTER].

TI-73   Press [2nd] [LIST] and then (▶) three times to go to the **CALC** submenu. Select LinReg(ax+b) or QuadReg. Press [ENTER].

Graphing

Calculator Functions: [STAT] and [LIST]

- Go to the **STAT** menu.

TI-83+   Press [STAT] and select EDIT....

TI-73   Press [LIST].

- Enter the input values into **L1** by entering the first number and pressing [ENTER], entering the second number and pressing [ENTER], and so on.

- Enter the output values into **L2** by pressing (▶), entering the first value and pressing [ENTER], entering the second value and pressing [ENTER], and so on.

Calculator Function: [PLOT]

- Press [2nd] and [Y=]. Select Plot1....

- Press [ENTER] to turn **Plot1** On.

- Select ⸭, ⟋, ⟘, ⬚, ⊘, or ⬚ under Type: by pressing (▼) once, (▶) repeatedly, then [ENTER].

Calculator Function: `Y=`

- Go into **Y= Editor**.

TI-83+   Press `Y=`. Enter the equation using `X,T,θ,n` as the variable. Press ⬇ to enter a second equation (if any).

TI-73   Press `Y=`. Enter the equation using `x` as the variable. Press ⬇ to enter a second equation (if any).

Calculator Function: `TABLE`

- To see the table of values for a function, press `2nd` and `GRAPH`. Use ⬇ or ⬆ to scroll through the table.

Calculator Function: `WINDOW`

- Set the graph with the domain and range that is appropriate to the scenario. Press `WINDOW`.

TI-83+   In `Xmin=` enter the lowest x-value and press `ENTER`, in `Xmax=` enter the highest x-value and press `ENTER`, and in `Xscl=` press `1` `ENTER`. In `Ymin=` enter the lowest y-value and press `ENTER`, in `Ymax=` enter the highest y-value and press `ENTER`, and in `Yscl=` press `1` `ENTER`.

TI-73   In `Xmin=` enter the lowest x-value and press `ENTER`, in `Xmax=` enter the highest x-value and then press `ENTER` twice, and in `Xscl=` press `1` `ENTER`. In `Ymin=` enter the lowest y-value and press `ENTER`, in `Xmax=` enter the highest y-value and press `ENTER`, and in `Yscl=` press `1` `ENTER`.

- Press `GRAPH` to see the function on the coordinate plane.

Calculator Function: [ZOOM] and [TRACE]

- Press [ZOOM].

- Select [ZStandard] for a standard coordinate plane, [ZoomStat] for statistical plots, or [ZSquare] for transformations.

- Press [TRACE] and (▶) or (◀). The screen should display **X=** and **Y=** at the bottom.

Calculator Function: [TRACE]

- Press [TRACE] and (▶) or (◀). The screen should display **X=** and **Y=** at the bottom.

SAXON™

A Harcourt Achieve Imprint